建筑与都市系列丛书｜世界建筑
Architecture and Urbanism Series | World Architecture

文筑国际 编译
Edited by CA-GROUP

Sustainability in Australia

澳大利亚：可持续性

中国建筑工业出版社

图书在版编目（CIP）数据

澳大利亚 : 可持续性 = Sustainability in Australia : 汉英对照 / 文筑国际 CA-GROUP 编译 . 一北 京 : 中国建筑工业出版社 , 2021.1
（建筑与都市系列丛书 . 世界建筑）
ISBN 978-7-112-25726-3

Ⅰ . ①澳 ... Ⅱ . ①文 ... Ⅲ . ①建筑艺术一介绍一澳大利亚一汉、英 Ⅳ . ① TU-866.11

中国版本图书馆 CIP 数据核字 (2020) 第 247372 号

责任编辑：毕凤鸣 刘文昕
版式设计：文筑国际
责任校对：王 烨

建筑与都市系列丛书｜ 世界建筑
Architecture and Urbanism Series ｜ World Architecture
澳大利亚：可持续性
Sustainability in Australia
文筑国际　编译
Edited by CA-GROUP
*
中国建筑工业出版社出版、发行（北京海淀三里河路 9 号）
各地新华书店、建筑书店经销
北京雅昌艺术印刷有限公司 制版、印刷
*
开本：787 毫米 ×1092 毫米　1/16　印张：14½　字数：460 千字
2024 年 7 月第一版　　2024 年 7 月第一次印刷
定价：**148. 00** 元
ISBN 978-7-112-25726-3
（36508）

a+u

建筑与都市系列丛书
Architecture and Urbanism Series

总策划 Production
国际建筑联盟 IAM　文筑国际 CA-GROUP

出品人 Publisher
马卫东 MA Weidong

总策划人/总监制 Executive Producer
马卫东 MA Weidong

内容担当 Editor in Charge
吴瑞香 WU Ruixiang

助理 Assistants
卢亭羽 LU Tingyu　杨 文 YANG Wen　杨紫薇 YANG Ziwei

翻译 Translators
英译中 Chinese Translation from English:
盛 洋 SHENG Yang (pp.16-p21, 114-119, 152-155)
熊赫男 XIONG He'nan (pp.22-113, 156-213)
日译中 Chinese Translation from Japanese:
吴瑞香 WU Ruixiang (p13) 陈旭丹 CHEN Xudan (p223)

书籍设计 Book Design
文筑国际 CA-GROUP

中日邦交正常化50周年纪念项目
The 50th Anniversary of the Normalization of China-Japan Diplomatic Relations

本系列丛书部分内容选自A+U第576号（2018年09月号）特辑
原版书名：
オーストラリアのサステイナビリティ
Sustainability in Australia

A+U Publishing Co., Ltd.
发行人/主编：吉田信之
副主编：横山圭
编辑：服部真吏　Sylvia Chen
摄影师：Fabian Ong（新加坡）
海外协助：侯 蕾

Part of this series is selected from the original a+u No. 576 (18:09),
the original title is:
オーストラリアのサステイナビリティ
Sustainability in Australia

A+U Publishing Co., Ltd.
Publisher / Chief Editor: Nobuyuki Yoshida
Senior Editor: Kei Yokoyama
Editorial Staff: Mari Hattori Sylvia Chen
Photographer: Fabian Ong (Singapore)
Oversea Assistant：HOU Lei

封面图：花园住宅，建筑设计：巴拉科与赖特建筑师事务所
封底图：唐纳森之家，建筑设计：格伦·马库特

本书第216页至229页内容由安藤忠雄建筑研究所提供，在此表示特别感谢。

Front cover: The Garden House designed by Baracco and Wright Architects. General view. Photo by Erieta Attali.
Back cover: Donaldson House designed by Glenn Murcutt. Atrium view. Photo by Anthony Browell.

The contents from pages 216 to 229 of this book were provided by Tadao Ando Architect & Associates. We would like to express our special thanks here.

Preface:

Why Australian Architecture?

RUAN Xing

Every now and then, a special issue magazine or a monograph on Australia's contemporary architecture is published in China. The Japanese, too, it seems, do the same, for this *A+U* special issue is not the first by this famed architectural publication to feature Australian architecture. Why Australian architecture? What exactly is its appeal to the Chinese (and the Japanese)?

The Chinese are not alone. The Welsh writer Jan Morris throughout her life was drawn to Australia (though not just to her buildings), experiencing a conundrum of love and hate for the country. This complex relationship culminated in her description of the affair as irresistible "distractions". Nearly two decades ago, Tsinghua University's *World Architecture* asked me to write an introduction to their special issue on Australian architecture. It was shortly after Glenn Murcutt received the Pritzker gong, and I thought such canonical recognition warranted an interpretation of the tenacious drive in Australian architects to move from their fringe position to the cultural centre. I thought, at the time (now I am somewhat pleased with proofs, such as this publication), that Australian architects deployed quite effectively three tactics to attract attention from New York, London and Tokyo, and to legitimise what they do. They are: myth reconstruction, larrikinism and doer worship.

Twenty years later, there has been no more Pritzker added to Australia's list of honours; the Japanese have since picked up a few more of the laurels. Of course, a Pritzker should not be used as a measure of a country's architectural accomplishments, but it is against this backdrop that I once again attempt to understand this new collection before us.

Certain threads that ran through my previous readings have, to my mind, continued in Australian architecture, but larrikinism is largely gone from this collection. Typified by a group of architects mainly based in Melbourne, larrikin and yet desperate techniques, such as deliberately distorting Corbusier and Aalto's building plans through hand-controlled motion photocopying, and then having them built... Some of these architects, though long past their years of "rebellious adolescence", are doing remarkably well – they are building large institutions and government buildings, and even remodelling the Sydney Opera House interior! But none of their works are featured here; neither judges from the cultural centre, nor the Japanese or the Chinese, take them seriously.

What we have before us are predominantly works heavily influenced by Glenn Murcutt and Richard Leplastrier (and they, too, are included in the collection). Most of the buildings are "elegant pavilions in the bush", though a few large urban projects, interestingly, led by the Germans and the French (Ingenhoven and Nouvel) with supporting local firms (not an unfamiliar story in China!) are also featured. The exquisitely, and in many cases expensively, made buildings nesting in the bush and sitting on the seashore are further embellished by their surroundings – the astonishingly beautiful Australian landscape. In the past two decades, the themes of "myth reconstruction of living in the bush" and "doer worship", two terms that I coined when I last introduced Australian architecture to Chinese readers, have been enthusiastically carried further by the students, and the students of the students of Murcutt and Leplastrier.

The works of Murcutt and Leplastrier, along with a few others such as Andresen (a Norwegian migrant herself) and O'Gorman, have been consistent in both themes. In their buildings, the "doer worship" of the resourceful and all-round Australian bushman is represented as a robustness

序言：

澳洲建筑解谜

阮昕

每隔一段时间，在中国就会有某个期刊或出版社为澳洲建筑出专集。日本人似乎亦乐于此道；a+u 为澳洲建筑出专辑也不是第一次。这个现象背后原由为何?澳洲建筑于国人（和日本人）究竟有什么吸引力?

其实国人对澳洲的兴趣并非孤例。威尔士作家简·莫里斯 (Jan Morris) 一生着迷于澳洲（倒不见得仅是对其建筑有兴趣），但却是爱恨交加的感受。莫里斯最终竟将这种复杂的心态描述为一种难于抗拒的“分心”。大约 20 年前，清华大学《世界建筑》期刊曾约稿，让笔者为其澳洲建筑专刊作序。那时正值格伦 · 马库特喜获普利兹克奖殊荣，我感觉应该接着如此主流权威对澳洲建筑师的认可，对澳洲建筑师从边缘地位迈向文化中心的立志作一番解读。当时笔者曾断言，澳洲建筑师之所以能令纽约、伦敦和东京刮目相看，是因为他们有效地启用了三个手段:“神话再造”“玩世不恭”和“崇拜实干”。现在看来，这本书对笔者当年的判断再次提供了部分佐证。

20 年后，澳洲建筑的光荣簿上没有再增添“普利兹克”，而“普利兹克”的奖簿上，日本建筑师倒是多了几位。诚然，我们不应该以普利兹克奖来衡量一个国家的建筑成就。不过这是一个我们试图理解眼前这本新书不可回避的背景。

笔者上次依赖于理解澳洲建筑的某些线索，现在看来，依然如故。但“玩世不恭”的手段在这本书里已不见踪影。曾经是一批扎营墨尔本的建筑师，他们通过看似玩世反讽，实而孤注一掷的手段，例如将柯布西耶和阿尔托的建筑平面，在复印时有意晃动，导致结果变形，然后将变了形的建筑平面直接盖成房子……这群建筑师中的成功者，如今早已过了“青春叛逆期”，在澳洲也已如鱼得水，拿到了很多大机构和政府项目，甚至还在做悉尼歌剧院室内改建！可是他们的作品在本书中完全没有出现。看来无论是传统意义上的文化中心，还是中国人或日本人，都没有把他们当回事儿。

这本书中的作品几乎都在不同方面深受马库特和理查德 · 莱普拉斯特里尔的影响 (他们自己的作品也选登在这本书里了)。大多刊登在这本集子中的建筑都是“丛林中秀丽的亭子”。当然也有几幢都市大建筑；只是它们是由德国和法国建筑师（英格霍芬和努维尔）和澳洲事务合作的结果（如此现象在中国早已是司空见惯)。这些精巧雅致的“亭子”建筑，或隐藏于丛林，或相伴于海岸，均得益于建筑周边美轮美奂的澳洲风景。近 20 年来，“重建丛林之居的神话”和“崇拜实用”这两个我曾经用来理解澳洲建筑的手法，又经马库特和莱普拉斯特里尔的学生，以及学生的学生之手，得到了发扬光大。

马库特和莱普拉斯特里尔，以及其他建筑师如安德森（挪威裔）和奥格曼都在践行以上两个主题。在他们的作品里，对澳洲早期殖民者那种因地制宜、足智多谋的实干精神致敬，体现在房屋的轻质框架结构的粗野感上，如暴露的钢木结构、全无天沟的瓦楞铁屋面（如同存放羊毛的仓库一般）。不过这些仅是表象：这些丛林小屋的空间组织，成熟的比例关系，以及马库特建筑中日渐舒适的室内环境，都印证了他们深受欧美现代建筑的影响，以及他们自身优越的“世界性”成长道路。如此“澳洲性”和都市的精炼成熟，在马库特和莱普拉斯特里尔的作品中往往有恰当的平衡。而在过去 20 年中，如本书所展现，他们的追随者们则是更加都市化，甚至如家养的宠物般，更加驯化了：过去边陲般的风范在他们的作品中变得风格化和精细化。在过度诚挚和不断精致化的驱使之下，我们目睹马库特的方法被照搬到悉尼市中心的大型办公

of lightweight construction, such as exposed timber and steel framework and the woolshed-like corrugated iron roofs without rain gutters. This is only the surface, for both the spatial organization and the sophisticated proportional control of their buildings, as well as an increasing level of interior comfort in Murcutt's houses reveal the deep influence of modern architecture and the authors' cosmopolitan upbringing. Such "Australian-ness" and urban refinement are well attuned. The works of their students in the past two decades (to some degree shown in this collection), however, are more urbane and even domesticated: the frontier character has been stylised and the construction refined. When the Murcutt approach is applied, with an earnest and increased refinement, to a large office building in central Sydney, the result is more like a gigantic piece of jewellery crafted with deadly perfection.

The works of Murcutt and Leplastrier have also been consistent in the sense that they continue to romanticise a way of living in harmonious coexistence with the landscape, and better still with minimal intrusion and damage to the land. Hence, we have the poetic maxims of buildings "touching the land lightly" from Murcutt; Leplastrier walks the talk and actually lives, with his partner and children, in the bush pavilion as seen in this book. Such a way of living would not be terribly convenient as far as a Chinese urbanite is concerned, especially when one has to drive to the nearest shop to buy a bottle of milk. However, most of these buildings are in fact weekend and holiday houses in an Australian bush where forest fires are frequent and destructive.

The theme of this selection is "sustainability". In a specific take on this heavily loaded problem, Murcutt and Leplastrier, and their students, are serious and skilful in making their buildings self-subsistent – to some extent off-the-grid. This is done, through skilfully designed building skins that are operable to absorb and release heat, as well as provisions of renewable energy generation, rain water harvesting, responsible waste management, and lightweight and prefabricated construction. It was around 2005, when I was at the helm of Architecture at Murcutt's alma mater, the University of New South Wales, I invited Murcutt and Wendy Lewin to join the university to teach the 3rd year design studio. Since then, Murcutt and Lewin, aided by a group of their enthusiastic followers, have been preaching just that to the younger generation.

Chinese readers ought to be reminded that most Australians live in the city, or more precisely in the sprawl of suburbs spreading out from city centres of concentrated business and workplaces. The honourable attempts of Murcutt, Leplastrier and the like in achieving a sustainable way of living represent a special kind of locality and luxury that can not possibly be replicated elsewhere. The principles (not the stylistic features), I am inclined to believe, can be and should be tested and used in large-scale urban buildings, for many of them are simply applications of age-old common sense. We see in this collection the most advanced construction of a layered glass façade that "breathes" to enable ventilation and energy preservation. But a glass building, as compared to a building of substantial thermal mass, is after all a glass building that absorbs and loses heat at the wrong times. Adapting the glass skin of a building to "breathe" or through plant cover only makes the building look "sustainable". This brings me to conclude this prologue by urging Chinese readers to see the problem of sustainability, through the lens of contemporary Australian architecture, not merely as a technical term or a stylistic feature, but as a moral obligation in a specific cultural context.

建筑上，其结果如同一个雕琢完美，而毫无生气的硕大首饰。

将一种与自然景观和谐相处的生活方式浪漫化，马库特和莱普拉斯特里尔也是一如既往。更为重要的是，他们力图使建筑对环境和土地的干预减到最小。于是，我们有了马库特耳熟能详的诗意口号："让建筑轻轻地触及大地。"在本书中读者可看到，莱普拉斯特里尔更加知行合一，自己和家人都住在丛林中"亭子"里。如此生活方式，如果连买瓶牛奶都需开车出行的话，中国的城市居民肯定觉得十分不便。不过，绝大多数的丛林之屋都是周末度假别墅，而毁灭性的山林火灾在澳洲是家常便饭。

本书的主题是"可持续性"。针对这个定义模糊的概念，马库特和莱普拉斯特里尔，以及他们的学徒们大多是严肃而技艺娴熟的；他力图将建筑做成一个自我维持状态。简言之，通过将房子的表皮做成巧妙而可开启调节状态，以便在需要时保温或散热。同时也会有不同程度的再生能源发电，雨水收集及环保排污处理和轻型预制结构，令房子基本上可以脱离市政管网。大约在 2005 年前后，笔者在掌理马库特母校悉尼新南威尔士大学建筑系时，特地邀请马库特和温蒂 · 卢因夫妇回校执教三年级设计课。自那时起，在几个热情追随者的辅佐下，马库特和卢因多年来一直给年轻学子传授以上设计思想和技能及技巧。

在此务必提醒中国读者，绝大多数澳洲人都住在城市里，或者更确切地说，是住在与商业办公区集中的市中心相接而延绵不断的郊区。马库特和莱普拉斯特里尔力图打造"可持续"的生活方式，其努力令人钦佩。但这是一个特殊场景里的一种奢侈品，不可在他时他处复制。而他们信奉的原则（而非风格样式），以笔者浅见，可以也应该应用于大尺度的都市建筑；其实这些设计原则都是经历了时间考验的常识。在这本书里我们可以找到高技的多层玻璃立面，令建筑表皮可以通过"呼吸"而保温或散热。然而，与一幢以"蓄热体"为主的坚实建筑相比，一幢透明玻璃楼毕竟是玻璃楼，其天性即是在不需要的时刻吸热和散热。通过高技将建筑的玻璃表皮变得可"呼吸"，或者干脆用植物将其遮挡，仅仅是使得建筑看上去像"可持续"而已。在此结束这篇短序，借鉴澳洲建筑这个他山之石，笔者希望敦促中国读者，务必将"可持续性"视为特定文化环境里的道义责任，而非仅是一个技术或风格的问题。

Ruan Xing
PhD, is Founding Dean and Guangqi Chair Professor of Architecture at School of Design, Shanghai Jiao Tong University. He was Associate Dean and Director of Architecture at UNSW Sydney (2004-2018), Head of School of Architecture at University of Technology Sydney (2002-2004).

阮昕
上海交通大学设计学院首任院长、光启讲席教授。曾任澳大利亚新南威尔士大学城市建筑环境学院副院长、建筑系主任（2004-2018），悉尼科技大学建筑学院院长（2002-2004）。

Sustainability in Australia

澳大利亚：可持续性

Editor's Word

编者的话

From infrastructures built by the indigenous tribe to the buildings made by local tradesman, the idea of sustainability is always seen responding closely to its climatic conditions, as well as cultural backgrounds. In Australia, these extremities found in the living conditions across the continent has provided the ground for unique architectural undertakings. As sustainable paradigms constantly shift the way we view and design our built environment, it is crucial that core traditions are not forgotten and new technologies are used purposefully with considerations given to its effectiveness and efficiency in its context. In this issue of a+u, together with Wendy Lewin, Maryam Gusheh and Tom Heneghan, the organisers of this year's exhibition, we bring to you a rich diversity of typology and styles of architecture, from its scale to its location, to introduce the notion of "sustainability" in the Australian context. (a+u)

从长久栖居于这片土地上的各民族的生活方式，到由当地地产开发者开发实现的建筑，澳大利亚关于可持续发展的理念，就常与气候条件、文化背景等密切相关。澳大利亚各地因地域不同而气候环境各异，我们因此从遍布全国的独特建筑中寻找到一些独特的案例。通过考察建设环境，再反映在设计上，我们得以继续发展"可持续发展"这一概念。新技术是为了使设计更加效率化的应用，是与地域文脉相适应，并非要舍弃传统。本书是我们以2018年举办的一次建筑展览会为契机，与温迪·莱文、玛丽亚姆·古什和汤姆·亨汉等作者一起合编的特辑，从多种建筑类型到建筑风格，从小住宅到大规模综合设施，介绍澳大利亚地域文脉之下大家对"可持续发展"的思考。

(a+u)

Project Mapping

项目分布图

01 **Shearers' Quarters and Captain Kelly's Cottage**, North Bruny Island, Tasmania
02 **The Garden House**, Western Port, Victoria
03 **Flexi-frame House**, Cirrus MVR Nomadic Water Conserving Bathroom, Mobile Hyper Kitchen Technology Demonstrator, Melbourne, Victoria
04 **Robe- trop_pods**, Robe, South Australia
05 **Anbinik- trop_pods**, Kakadu National Park, Northern Territory
06 **Mooloomba House**, North Stradbroke Island, Queensland
07 **Lovett Bay House**, Northern Beaches, New South Wales
08 **Donaldson House**, Pittwater, New South Wales
09 **Australian Opal Centre**, Lightning Ridge, New South Wales
10 **Cranbrook School Wolgan Valley Campus**, Greater Blue Mountains National Park, New South Wales
11 **Les Wilson Barramundi Discovery Centre**, Karumba, Queensland
12 **East Pilbara Arts Centre, Newman**, Western Australia
13 **The Ian Potter National Conservatory**, Sydney, New South Wales
14 **1 Bligh Street, Canberra**, Australian Capital Territory
15 **International House Sydney**, Sydney, New South Wales
16 **The Brewery Yard**, Central Park, Chippendale, New South Wales
17 **One Central Park**, Chippendale, New South Wales
18-1 **Water Urbanism in Australian Cities**, Melbourne
18-2 **Water Urbanism in Australian Cities**, Brisbane
18-3 **Water Urbanism in Australian Cities**, Perth

Climatic Zones Diagram／气候带（示意）图

- Equitorial／赤道气候
- Tropical／热带雨林气候
- Sub-Tropical／亚热带气候
- Desert／沙漠气候
- Grassland／热带气候
- Temperate／温带气候

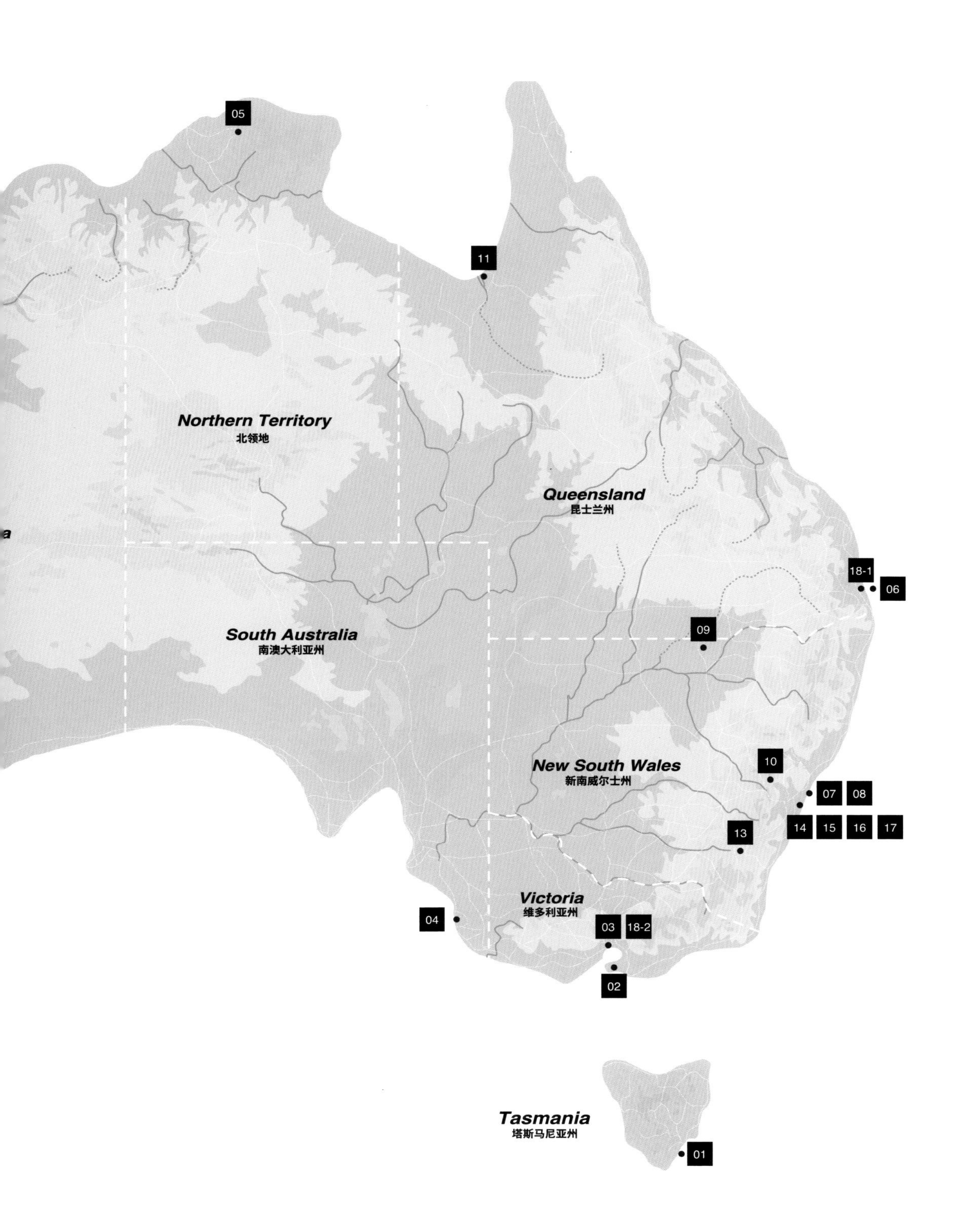

05
11
Northern Territory
北领地
Queensland
昆士兰州
18-1
06
South Australia
南澳大利亚州
09
New South Wales
新南威尔士州
10
07
08
14
15
16
17
13
Victoria
维多利亚州
04
03
18-2
02
Tasmania
塔斯马尼亚州
01

Interview: Wendy Lewin

Sustainability in Australia, and *Universal Principles: Unique Projects* Exhibition

访谈：温蒂·卢因

澳大利亚的可持续性及“普遍原理：特殊项目”展览

a+u: Thank you for taking time down to do this interview with us. The last time a+u featured architecture in Australia was more than 10 years go, and we did the topic "Living off the land", which featured mostly residential houses. Hence, we felt it is time we revisit the architecture scene in Australia. Coincidentally, you had been working on the ongoing exhibition at Tokyo City View in Roppongi, Universal Principles: *Unique Projects - Australian Architecture Resetting the Agenda*, in conjunction with Australia now. So, we felt it was really appropriate to have you on board with us for this book which focuses on sustainability in Australia. Perhaps, you would like share with us more about the exhibition title, Universal Principles: *Unique Projects*.

Wendy Lewin (WL): By way of a very brief background to the exhibition and its title, initially there was an expressed interest from the organizing committee in Tokyo that the exhibition's focus be on the experience Australian architects have in the design of "sustainable" architecture and their response to "nature".

I suggested that perhaps the focus of discussion and practice in Australia can be seen to have broadened or developed in relation to these interests, that "nature" is rarely considered "scenographically" or separately. Rather that place, context and nature – the true public realm – are increasingly *fundamentals* in our briefs, that very different architectures are evolving in Australia as we seek solutions that are congruent and supportive of nature, context and resources across the distinctly different regions of our continent.

Architects, engineers, our clients, commissioning agents, many developers and authorities have for years now sought to engage with more site responsive and responsible development strategies – strategies that often incorporate sophisticated takes on first principle technologies to positively address specific contextual and environmental issues. There is, for me, a true intellectual and aesthetic elegance in the way that ancient technologies and systems are informing the most advanced and increasingly – importantly – guiding contemporary urban and non-urban architecture.

The exhibition provided a unique opportunity to present and discuss a range of architectural projects that are distinct, projects that would significantly broaden the understanding of Australian architectural practice, projects that would reveal the relationship of Australian architects to their culture and a collective ambition to address contemporary social and environmental challenges.

So going back to the exhibition title: I intended that "Universal Principles" refers to and respects the global history of architectural ideas, their relevance and inventive applications as well as the cultural and environmental continuums within which we practice. "Unique Projects" references the personal ways in which architects consider or interrogate their commissions and shape their work, their engagement in creative collaborations and their particular interests in addressing the specifics of place / context. It also alludes to the unexpected.

a+u：感谢您抽出时间接受我们的采访。上一次《a+u》策划澳大利亚建筑专题，我们做了“远离尘嚣的住宅”主题(a+u 07:08;a+u China 08:06)，主要关注住宅建筑。但那已经过去了十多年，感觉是时候重访澳大利亚的建筑了。巧合的是，您之前也参与准备了在东京城市观景“天空画廊”举办的展览“普遍原理：特殊项目——澳大利亚建筑的重启”。所以，这次能有您与我们共同聚焦澳大利亚的可持续性，可以说再合适不过。那么，您愿意先和我们分享一下“普遍原理：特殊项目”这个展览题目背后的故事吗？

温蒂·卢因：这个展览及其标题的背景，简而言之，最初是源于东京组委会的浓厚兴趣。他们希望将澳大利亚建筑师对“可持续”建筑的设计和对“自然”的回应作为展览焦点。

应该可以看到，在澳大利亚，讨论和实践的焦点已经扩散，或与环保建立起了关联，这使得“自然”很少被视为一种“布景”或独立的存在。相反，场所、环境、自然，诸如此类真正的公共领域越来越重要。我们寻找不同方案，以适应各地独特的自然、环境、资源状况，而与此同时，澳大利亚的建筑形态也趋于多样。

许多年来，建筑师、工程师、我们的客户和委托方，以及众多开发商和政府机构都在寻找更具适应性、更负责任的地块开发策略，基于“第一原理”调整复杂的技术措施，积极应对特定的文脉和环境问题。对我而言，古老的技术和系统中存在着一种真正的智识和典雅的审美，由此启发最先进的建筑理念，并在当代都市建筑和非都市建筑的实践中起到越来越重要的指导作用。这个展览为展示、讨论一系列不同的建筑项目提供了独一无二的机会。这些项目可以极大地拓展人们对澳大利亚建筑实践的理解，揭示澳大利亚建筑师及其自身文化之间的关系，以及他们应对当代社会和环境挑战的雄心。

回到展览的标题，我倾向认为，“普遍原理”指的是建筑思想全史，展现各建筑理念的相关性和革新性，说明我们所置身的文化、环境的延续性。“特殊项目”则是指建筑师个体如何审视他们接到的委托，如何打造他们的作品，如何参与创造性的合作，以及关注场所、文脉的哪一部分；当然，也包含了一些打破常规的做法。

a+u：那么，您是怎样决定最终展出的12个项目的？

温蒂·卢因：这12个项目的选择参考了一个基本标准，即每个案例都必须是所在领域的范例，能够催生一个进步的未来。设计者不是肤浅地应用环境技术和措施，而是赋予作品独特的个性，使之与场所或环境契合。这些项目必须通俗易懂，让人无需借助讲述和联想的方式便能理解。它们不能用复杂的外观来掩盖单调的处理方式，不能只是依赖环境技术，也不能误读甚至脱离自身的背景。应东京组委会成员的要求，澳大利亚新南威尔士州闪电岭的澳大利亚欧泊中心(见108-113页)被纳入展览。作为未实现的项目，它像是一篇烘托展览主旨的“个人主张”，提供了一个自主性建筑的案例，展现在澳大利亚半干旱地区某个复杂的文化项目如何自给自足。而我寻找的出展项目，尽管规模、背景(城市或非城市)各不相同，但都能够巧妙地应用技术，实现高水平的设计，给讨论未来发展提供新的视角

a+u: So, how did you decide on the 12 projects for this exhibition?

WL: The 12 projects were selected on one fundamental criteria. Each case must be an exemplary work of architecture in its own right and be shaping a progressive future. In addition they had to be projects in which environmental technologies and solutions are not retrofitted or superficially applied but rather have caused for an inventive poetic to inhabit the work and directly reference the particularities of the given place or context. The projects had to be clear and legible, able to be interrogated and understood without the aid of narratives and metaphors. They could not be reliant on screens and camouflage to distract from an underlying, prosaic resolution of their programs, applications of environmental technologies or a misreading of, or detachment from, their context. The inclusion of the Australian Opal Centre (See pp.108–113) at Lightning Ridge was at the request of the Tokyo organizing committee members. As an unrealized project it served as a 'personal thesis' on which to anchor the theme of the exhibition and offers an example of autonomous architecture that supports a complex cultural program in a semi-arid region of Australia. I sought projects which, through their varying scales, urban and non-urban contexts and programs could cause for the intelligent application of technologies and elevated levels of design, to reset or refocus the discussion around future development.

a+u: That takes me back, so what does sustainability mean to Australians?

WL: It means many different things and I can only offer my thoughts. I began my architectural education in the early 70s when the notion of protecting "spaceship earth" and the "greening of the planet" were the *causes célèbres* ... and they needed to be, they still need to be. These challenges at the time helped to form and focus global debate, discussions and legislation around "sustainable" practices, energy, land use and so on that we benefit from today. Ironically since then, relatively stable global political, social and financial circumstances have rendered sustainability and sustainable "umbrella terms" - loosely applied to a multitude of practices or activities that are actually neither. For me the notion of sustainability is an important, valued and sadly unachievable ideal, only because it's implausible to think that anything, any resource can be continuously drawn on - that it won't run out. Even in relation to the use of timber, globally we are creating greater and greater areas of dedicated plantation timber forests, producing monocultures of species with limited, commercially shortened life cycles. In nature, biodiversity in flora and fauna requires complexity, seasonal nuance and the passage of time to be healthy, and we are affecting this badly, dramatically. I suspect throughout the world "sustainable" and "sustainability" are most commonly the branding terms now used to support and promote greater consumption of goods, services, new areas of land released for development and so on. Why not think about how responsibly and resourcefully you can work with what you have and consciously work to minimize the impact of your occupation – as architects take charge of shaping this much needed progressive agenda?

a+u: Is there any point in history or cultural part of Australia that influence the way people think and live in Australia?

WL: I'm sure there have been many such moments in time, the social, cultural and political identity of Australia in the past has been, let's say, "fluid". Australia has an enviable history of cultural, technological and investigative research and achievements, all of which I suspect were driven by necessity, blind optimism and the luxury of creative freedom. And contemporary Australia is a nation of many nations.We all benefit from the diverse opportunities that multiculturalism fosters and the fundamental stability of our system of government. We fall short on achieving equality and conciliation in all areas of life. My sense is that anyone who has lived or lives in Australia has had to appreciate it on their own terms and over time understand it's a country that reveals itself in bits. It's not the sum of the clichés that are offered to the world.

a+u：话说回来，对澳大利亚人而言，可持续性意味着什么？

温蒂·卢因：它意味着许多，在这里我只能表达我的想法。我1970年代初开始学习建筑时，正值保护“宇宙飞船地球号”和“绿化地球”的风潮涌起……这并不奇怪，而且放在今天仍有必要。也恰恰是这些挑战，在当时引发了全球关于可持续实践、能源、土地使用等问题的讨论，并加速了相关法制环境的形成，使得我们今天能够受益。然而，具有讽刺意味的是，从那时起，相对稳定的全球政治、社会和金融环境，导致“可持续性”和“可持续”成为“伞式术语”（译注：指如同大伞一般涵盖多术语的总括型术语），被滥用于事实上二者皆非的实践。对我来说，“可持续”是一个重要的、有价值的但很遗憾无法实现的理想概念，原因就在于，那种认为任何东西、任何资源都可以不断被利用——仿佛不会耗尽——的想法是不切实际的。以木材使用为例，如今我们在全球范围进行越来越大规模的树木种植，培育那些为商业目的缩短生命周期的有限物种。但自然界有着自身的复杂性、季节差异性和时间积累，以保持动植物的多样性。可以说，人类的行为正在极大地破坏这一规律。在我看来，如今在世界各地，为了支撑和促进更多的商品和服务消费、释放新开发用地等等，“可持续”和“可持续性”都已经成了最常见的宣传用语。为什么不想想怎样负责任地物尽其用，尽可能减轻人类行为的影响？作为建筑师，或许你可以担起重任，推动这一迫在眉睫的革新。

a+u：澳大利亚当地人思考、生活的方式，有没有受到某些历史或文化的影响？

温蒂·卢因：肯定有许多这样的时刻。一直以来，澳大利亚的社会、文化和政治特性都可以说是“流动的”。这里的文化、技术研究及其成果源源不断，令人艳羡；我认为这固然是必然的趋势，却也有赖于我们无条件的乐观主义和自由创作的氛围。当代的澳大利亚是一个多民族国家。我们所有人都受惠于多元文化带来的各种机会，得益于政治体制的基本稳定。但与此同时，我们很少实现生活中的平等或融合。我的感觉是，凡在澳大利亚生活过、居住过的人，都要用自己的眼光来欣赏它，并在时间的推移中明白，这是一个碎片化的国家。那些流布坊间对澳大利亚的既定评价，都不足为信。

a+u：的确，我们的人生观因环境而异。我还注意到有些项目被授予了可持续性评级。您对澳大利亚的可持续性评级系统怎么看？我感觉某种程度上，所有人都在追赶这一潮流，以至于评级变成了一种营销手段。就澳大利亚来说，这种评级对建筑真的有利吗？

温蒂·卢因：没有定论。评级的存在很可能有好处，但我认为它们本身并不重要。当然，对于那些“斗志昂扬”且从中攫取经济利益的人来说，这很有吸引力，毕竟能够借此宣扬某种“竞争优势”，而且自我贡献也得到了认可。但重点是，你在哪里划线。几星是有意义的？一星、半星还有价值吗？就像越来越多的行业奖项一样，这些星级很大程度上就是品牌和营销。最近一位观察敏锐的同事注意到，我们几乎步入了这样的时代：某个重量级建筑奖项的设立，可能只是为了表彰一座建筑的落成！

本次展览的项目都超出了此类评级的标准，它们代表着高密度城市环境下，创造性实践的巨大发展。由于本身的自主性，这些项目可以不依赖城市提供的服务而独立存在。它们对城市的部分垃圾和水进行再处理，其中一个案例还能为邻近的开发项目提供电力。大规模热交换系统的应用、三重热电联产发电厂的引入、大面积的光伏阵列、灰水和黑水的处理及循环系统，反映了开发商、立法者和顾问的多方协作，这些都远远超出了“星级”标准。又或许评级系统对这些雄心勃勃的项目起到了促进作用？谁知道呢。打造符合评级标准的作品，在经济上是明智的……但如果我们在推进城市密集化的同时，还打算改善城市环境和市民生活，那些有创造性、对环境高度负责、有变革性的项目就会产生长远的效益。

a+u：目前澳大利亚建筑教育的发展态势如何？学生们的兴趣和教学方法是怎样的？

温蒂·卢因：目前，学生数量呈上升趋势，希望这意味着人们对于建筑学的潜力和相关性更感兴趣了。不过，这门学科仍在变化。如今建筑学的教学方式与几十年前已大不相同，许多教育机构，包括一些大学，都在建筑大类下开设更通用的学士学位，将学科重点扩展至项目管理、室内建筑、景观建筑、应急建筑等领域。学生人数不少，直接接触的

a+u: Indeed, there are different aspects of how we live depending on the environment. I noticed there are a few projects that have obtained sustainable ratings. What do you think about sustainable rating tools for in Australia? I feel that at some point, everyone was jumping on this bandwagon, so much so it became a marketing tool. In the case for Australia, do they bring any benefit to its architecture?

WL: It's arguable. There are probably benefits to having ratings but I think in themselves they're not significant. Of course celebrating some sort of "competitive edge" and recognition of contribution is always attractive to those who identify "aspirationally" and benefit economically from such things. But surely there has to be point where you draw the line. How many stars are meaningful, and actually what is the value of a star ... a half star ... really? Like the ever increasing number of industry awards, yes, stars are very much about branding and marketing. An insightful colleague recently observed that we are almost at the point in time when a named architectural award will be created for a building just because 'it stands up'!

The projects in the exhibition assessed with the star ratings tool exceed the criteria for such ratings and represent a significant development in the creative trajectory of practices working on projects in high-density urban environments. Autonomous in their own right they can exist within the city independently without city supplied services. They are reprocessing some of the city's waste and water and in one case supplying power to adjacent developments. The application of large-scale heat exchange systems, the inclusion of trigeneration power plants, extensive photovoltaic arrays, and grey and black water treatment and recycling systems reflect the commitment of developers, legislators and consultant collaborative to going far beyond the criteria for "stars". Who knows, maybe the star rating systems was formative in these ambitious projects? It is financially smart to produce work that accords with these rating systems ... and creative, highly responsible and inventive works will provide benefits in the future as we find ways to achieve higher densities in our cities and at the same time significantly improve the environment and lives of people.

a+u: With regards to architecture education in Australia, what is the situation like there? Perhaps, in terms of students' interests and teaching methods?

WL: Currently, the number of students are increasing, hopefully that reflects greater interest in the potential, relevance of architecture - but the discipline is changing. The way architecture is taught now is very different from how it was a couple of decades years ago. Many institutions are moving towards a more generalized first degree with some universities offering under the umbrella of Architecture, expanded areas of disciplinary focus such as project management, interior architecture, landscape architecture, emergency architecture and so on. Direct contact teaching hours have radically reduced, student numbers are high, there is a sense that we are not equipping students with much needed skills and interest for practice ... perhaps that's always been the case ... perhaps it's only practice that 'equips'.

a+u: What do you think about the future of the architecture industry in Australia? What are the challenges to be faced?

WL: All levels of government are showing greater interest in making better cities, supporting technological innovation and autonomy and developing better and more responsible land use - not in themselves new goals but increasingly, creative projects in these arenas are being premiated which in turn broadens the opportunities for engagement within the profession. In terms of challenges, the projects that you see in the exhibition, the research and development that supported many of them, "took on the battle". The tenacity and creative work of these architects, their clients and their collaborators support my sense that we will produce more relevant, well-judged and worthwhile work. I'm far more positive about the future than I may have been 10 or 15 years ago. I would like to think that these projects show the strategic value of environmentally responsible and responsive considerations, the creative application of technologies and the cultural value of elevated design.

Interviewed on 4 July 2018,
Australia Embassy Tokyo.

学课时却骤减，我们不确定是否赋予了学生应有的技能和实践兴趣……也许事实一直如此……也许只有实践才能做到这些。

a+u：您怎样看待澳大利亚建筑业的未来？它将面临哪些挑战？

温蒂·卢因：现在各级政府部门更关注的都是如何营造更美好的城市，如何支持技术创新和区域自治，如何实现更加合理有效的土地开发。这些目标本身并不新鲜，但已有越来越多的创意项目在相关领域浮现，反过来也给建筑师提供了更多的参与机会。在挑战方面，你在展览中看到的许多项目，其背后的研发都堪称“迎难而上”。这些建筑师及其客户、合作者的坚持不懈，还有他们创造性的工作，都让我感觉，我们将孕育出更多考虑周全、意义深远的作品。比起10年或15年前，现在我对未来的看法积极得多。我相信，这些项目显示出了对环境敏感且负责的战略价值、技术的创造性应用以及深度设计的文化价值。

（采访于2018年7月4日，澳大利亚驻东京大使馆）

Student Contribution: Drawings
The drawings for this publication are from the exhibition, Universal Principles: Unique Projects – Australian Architecture Resetting the Agenda. Produced by students of architecture based in Sydney, with direct reference to documents provided by associated practices.

学生供稿：图纸
图纸来自展览“普遍原理：特殊项目——澳大利亚建筑的重启”，由悉尼的建筑系学生制作，直接引用自相关的实践文本。

Students: Jason Cheung (Conceptual and Prototyped House and Service Units), Farros Ghozi Djojodihardjo, Janice Ma (International House Sydney), Natalie Wing Sum Ho, Jincheng Jiang, Yue Yin (Shearers' Quarters and Captain Kelly's Cottage), Le Thao Nguyen Nguyen (One Central Park), Yishun Tang (Lovett Bay House)
Drawing Development and Refinement: Rebecca Fray, Natalie Wing Sum Ho, Jincheng Jiang
Drawings Architecture Program, University of New South Wales, Sydney led by Maryam Gusheh, Wendy Lewin, William Maynard

Wendy Lewin is a practising Australian architect and educator at the University of New South Wales, Sydney.
温蒂·卢因，澳大利亚执业建筑师，悉尼新南威尔士大学教师。

John Wardle Architects

Shearers' Quarters and Captain Kelly's Cottage

Waterview, North Bruny Island, Tasmania 2011, 2016

约翰·沃德尔建筑师事务所

剪羊毛工人的宿舍和凯利船长的小屋

塔斯马尼亚州，北布鲁尼岛，沃特维尤 2011，2016

Shearer's Quarters

剪羊毛工人的宿舍

The Shearers' Quarters is located on Waterview, an historic farm on North Bruny Island, first granted to Captain James Kelly in 1840. The 440-hectare property remains a working sheep farm. It has been operated by the Wardle family for 15 years, with the rejuvenation of the landscape a core priority. Occupying the site of the old shearing shed, destroyed by fire in the 1980s, the project is a companion building to the existing historic cottage, renovated in 2016. It now accommodates shearers, the family, friends, and John Wardle Architects' staff on annual retreats.

An exposition of two primary forms of Australian agricultural structures, the plan form transitions along its length to shift the profile of a slender skillion roof at the western end to a broad gable at the east, referencing both the fall of the land to the south and the line of the original residence along its north. Whilst the stone retaining walls were constructed by a stonemason on the island, using stones sourced from the property, the relatively remote island site required all other construction materials be transported via ferry from the mainland. This significantly determined the project's lightweight construction strategy.

The material palette is singular: corrugated galvanized iron to the exterior, and timber internally. The primary internal lining is Pinus Macrocarpa, sourced from many different suppliers, principally as individual trees from old rural windbreaks. The bedrooms are lined in recycled apple-box crates, sourced from the many old orchards of the Huon Valley, where the timber has remained stacked but unused since the late 1960s. The flooring is recycled Yellow Stringybark. The east chimney, in old handmade clay bricks, references three original fireplaces that are similarly composed on the western external walls of the original dwelling. The planar and volumetric shifts — operable wall elements that mediate breeze, light and infuse the interior with the local atmosphere — recall the beauty, economy and inventiveness of 18th-century transformation furniture.

In this refined architectural work, generic building types and attendant systems are adjusted to the intricacies of the local setting, encompassing historical, material and environmental parameters.

pp. 22–23: Shearers' Quarters cladded in galvanised, corrugated iron, is positioned on the site of the old shearing shed alongside Captain Kelly's Cottage cladded in timber. Opposite: The roof is angled to help channel excess rainwater into the nearby river. All images on pp. 22–43 by Trevor Mein.

第22-23页：剪羊毛工人的宿舍位于剪毛棚旧址上，外立面由波纹状镀锌铁板包裹，与由木料包裹的凯利船长的小屋比肩而立。对页：屋顶倾斜有助于将多余的雨水引流到附近的河流中。

剪羊毛工人的宿舍位于北布鲁尼岛上一个历史悠久的牧场沃特维尤，这片牧场于 1840 年首次被授予詹姆斯·凯利船长。它占地 440 公顷，至今依然是一个运营中的绵羊牧场。以该地的景观重塑为核心及首要任务，这里已经由沃德尔家族经营达 15 年。本项目位于 1980 年代被大火烧毁的原剪毛棚旧址上，作为富有历史意义的船长小屋的配套建筑，于 2016 年被翻新。如今，这里为牧场的剪羊毛工人、沃德尔家族及其友人提供住所，同时也是约翰·沃德尔建筑师事务所员工们每年度假的地方。

为了同时表现澳大利亚农牧业用房两种主要的形态，该设计沿其面宽逐渐变化：从西侧末端细长的单坡屋顶轮廓过渡为东边宽阔的山墙。这样的设计既源自向南陷落的地形，也参考了这里原有住宅面北的轮廓线。尽管石砌挡土墙是由岛上的一个石匠用从本地开采的石头所建造的，但由于该项目地处较为偏远的岛上，所以其他建筑材料都需要从内陆通过渡轮运输至此地。这使轻量化建造策略成为必然选择。

材料的选用无与伦比：外部是波纹状镀锌薄钢板，内部是木材。室内基础衬板选用了大果松木，由多家不同的供应商提供，大部分都出自古老乡村防风林中的树木。除此之外，卧室内衬板材来源于对旧苹果板条箱的再利用，这些板条箱出自休恩河谷的众多老果园，它们自 1960 年代后期以来一直堆放在那里，无人问津。地面采用回收再利用的桉木铺装。东边的烟囱用从前的手工黏土砖砌成，参照了原来住宅西侧外墙上三个同样以此种砖砌成的原始壁炉。平面和体量的变化通过活动隔墙构件实现，它调节着气流和光照，并为内部空间注入当地氛围，重现了 18 世纪可变形家具的魅力、经济实用性和独创性。

在这件精致的建筑作品中，通用建筑类型和相关系统被加以调整，以适应当地历史、材料和环境的复杂性。

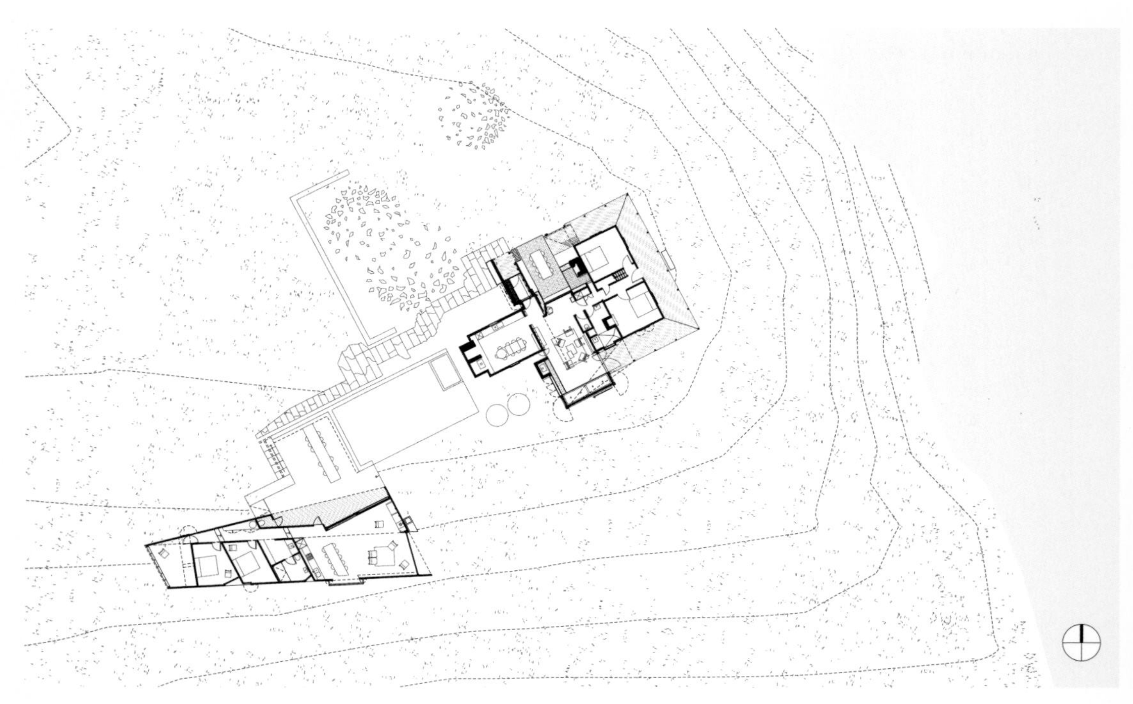

Site plan (scale: 1/700)／总平面图（比例：1/700）

This page: Openable vents and louvers allow for natural cross ventilation during summer.

本页：在夏季，可实现自然通风的开放型通风孔和百叶窗。

pp. 28–29: Shearers Quarters sits on a historic sheep farming property. pp. 30–31: View of the open plan living room lined with pine wood. This page: View of bedroom with walls covered with recycled apple crates from nearby orchards.

第28-29页：剪羊毛工人的宿舍位于这片历史悠久的绵羊牧场内。第30-31页：天花板和墙壁皆为松木饰面的开放式起居室内部。本页：以回收利用的苹果板条箱作为墙面材质的卧室内部。

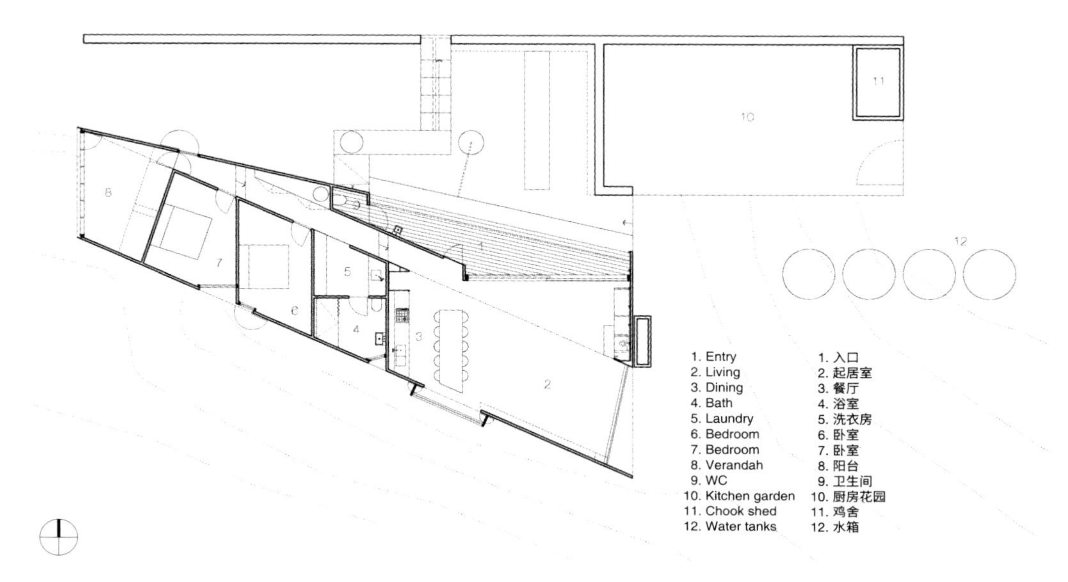

Floor plan (scale: 1/300)／平面图(比例:1/300)

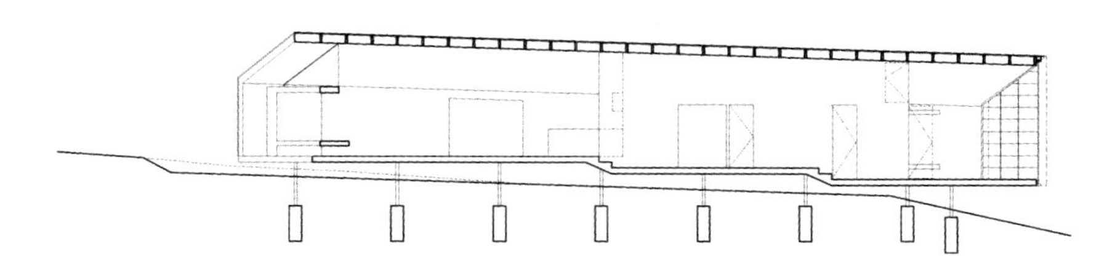

Long section (scale: 1/300)／纵向剖面图(比例:1/300)

Credits and Data

Project title: Shearers' Quarters
Client: Susan and John Wardle
Location: Waterview, North Bruny Island, Tasmania, Australia
Design: 2008
Completion: 2011
Architect: John Wardle Architects
Design Team: Andrew Wong, Chloe Lanser, Jeff Arnold
Consultants: Cordwell Lane (Construction), Gandy and Roberts Consulting Engineers (Structural Engineer), Holdfast Building Surveyors (Building Surveyor)
Project area: 136 m²

Captain Kelly's Cottage

凯利船长的小屋

pp. 34–35: Exterior view of the restored Captain Kelly's Cottage built in the 1830s. This page: View of one of the two existing structures that houses the bedrooms. Opposite: The verandah extends and encapsulates the new works, tying the new and old together.

第34-35页：建于1830年代的凯利船长的小屋被修复后的外观。本页：两座原有房屋之一的外观，该座被作为卧室使用。对页：游廊使空间扩大并将新设计囊括在内，新与旧得以融合在一起。

A companion building to the Shearers' Quarters, Captain Kelly's Cottage is a restoration of an existing weatherboard cottage. The original building had aged; it had deteriorated due to weathering and had been diminished by several unsympathetic alterations. The new work involved the intricate restoration of the original form, with added layers carefully removed. Unique construction techniques were revealed in the process and selectively exposed and emphasized. The original verandah inspired the contemporary architectural intervention — a new living area placed between the two existing structures, the kitchen and bedrooms. The layered soffit and exposed ceiling rafters were continued in the new entry and living spaces to connect new with old.

A sheltered north-facing courtyard is nestled into the leeward side of the cottage. A new chimney recalls the original — but long removed — chimney. Constructed of prototyped custom made white 'ghost' bricks, the reinstated chimney helps to both define and heat the courtyard space. The courtyard is located around the existing walnut and mulberry trees, both over 100 years old and as powerfully present as the cottage, which has acted as sentinel on the precipice of this remote location for more than 175 years. A dry-stone wall blurs the boundaries of the distant past and the present.

Credits and Data

Project title: Captain Kelly's Cottage
Client: Susan and John Wardle
Location: Waterview, North Bruny Island, Tasmania, Australia
Design: 2015
Completion: 2016
Architect: John Wardle Architects
Design Team: Andrew Wong and Danielle Peck
Consultants: Cordwell Lane (Construction), Gandy and Roberts Consulting Engineers (Structural Engineer), Holdfast Building Surveyors (Building Surveyor), John Matthews (Architectural Historian)
Project area: 320 m^2

pp. 38–39: Multi-paned windows can slide away to reveal a view of the landscape and external screens can fold across to hunker the house down during a storm. Opposite: View of the existing kitchen volume with its original brick formwork is now reinstated as both a kitchen and dining room.

第38-39页：多窗格的窗户被完全推开后可尽观风景；当暴风雨来临时，外部挡风屏可以展开并环抱整个房屋，使内部保持平静和舒适。对页：起初砖砌构筑的厨房被修复为厨房和餐厅空间。

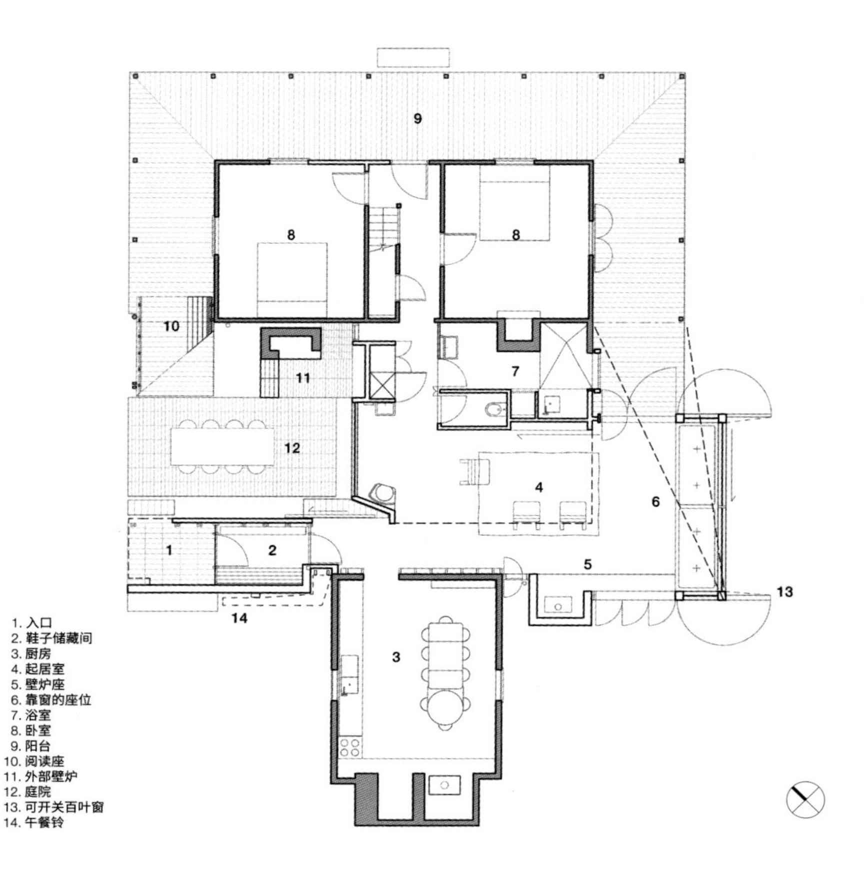

Floor Plan (scale: 1/200)／平面图(比例:1/200)

作为剪羊毛工人的宿舍的配套建筑，凯利船长的小屋是对原有的挡风小屋的一次整修。原建筑物已经老化，由于风化作用而变得萧条破败；加之此前几次与原设计不同的改造，使该建筑早已不复往昔。新设计包含对小屋原始形态复杂精细的修复，且须将后加的铺设小心地拆除。独一无二的建造技术在施工过程中尽显其能，这些技术被有选择地公开给观者并加以强调。最初的游廊为当代建筑的介入注入了灵感——在两个原有结构（厨房和卧室）之间设置一个新的起居空间。分层的挑檐内衬和裸露的屋顶椽子一直延伸到新的入口和起居空间，以此在新与旧之间建立连接。

有顶棚的庭院面向北，坐落在小屋的背风侧。庭院中新的烟囱使人不禁追忆起早已被拆除的旧烟囱。建造新烟囱所用的白色砖块是循着记忆，依照原型定制的，复原后的烟囱更好地定义了庭院空间，并使其成为一处温暖的存在。庭院围绕着场地上已有 100 多年历史的核桃树和桑树，它们有着与小屋相同的强烈存在感。小屋如岗哨般伫立在这个偏远地区的悬崖上已有超过 175 年的历史，干砌石墙的存在使过去与当下的界限变得模糊。

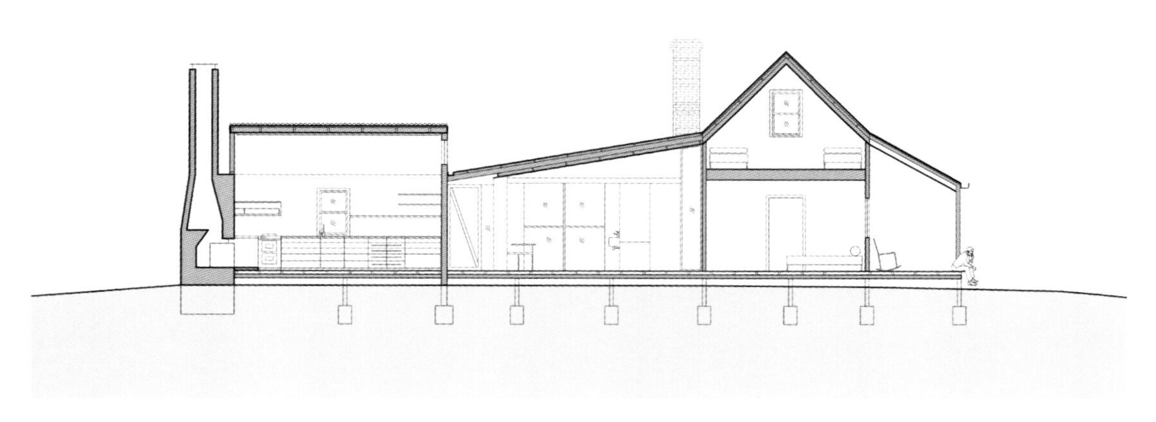

Long section (scale: 1/200)／纵向剖面图（比例：1/200）

This page: Interior view of the existing bedroom volume where the new additional elements heal the old.

本页：在原来的卧室空间内部，引入的新元素使老旧的空间获得重生。

Baracco and Wright Architects

The Garden House

Western Port, Victoria 2015

巴拉科与赖特建筑师事务所

花园住宅

维多利亚州，西港 2015

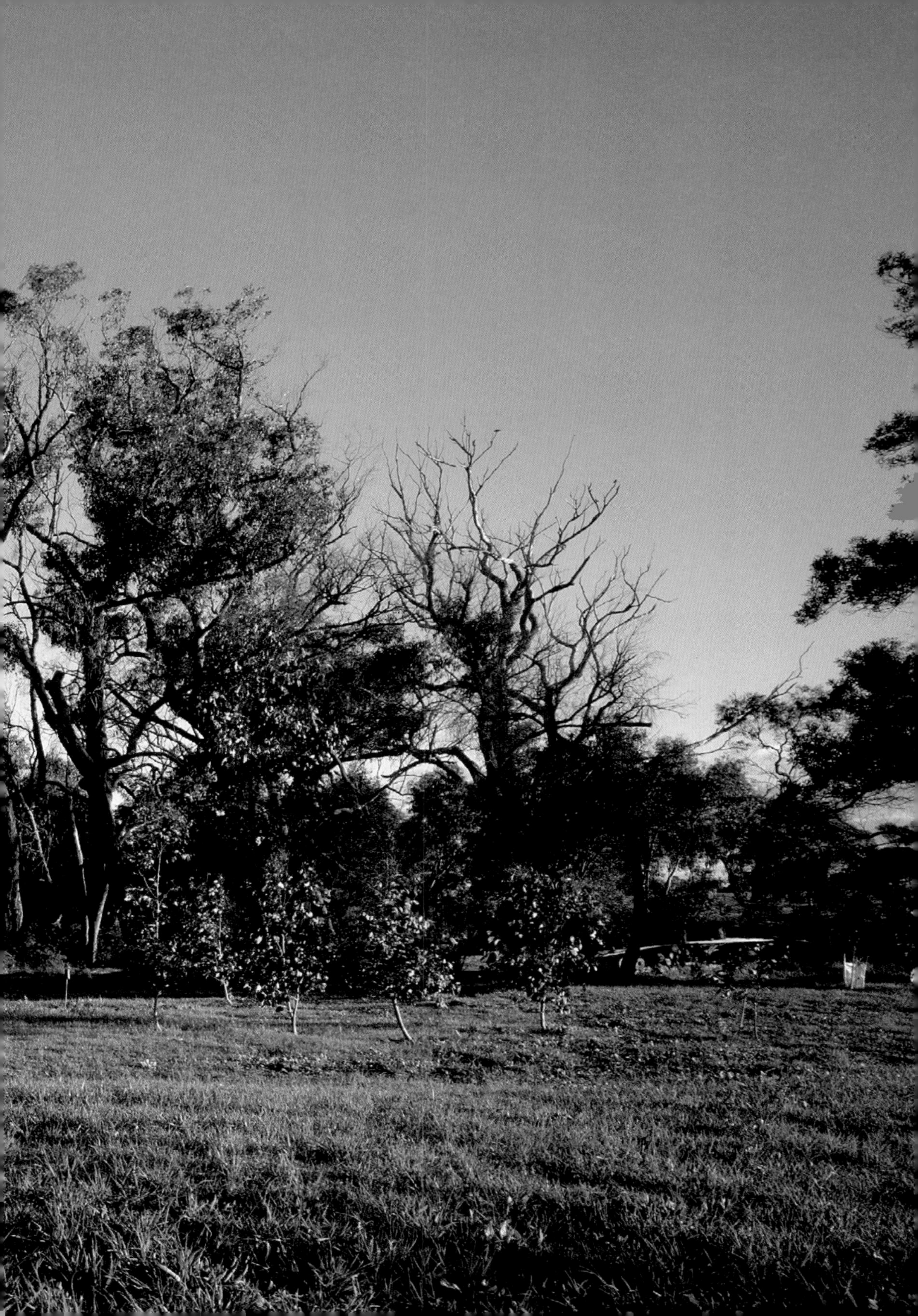

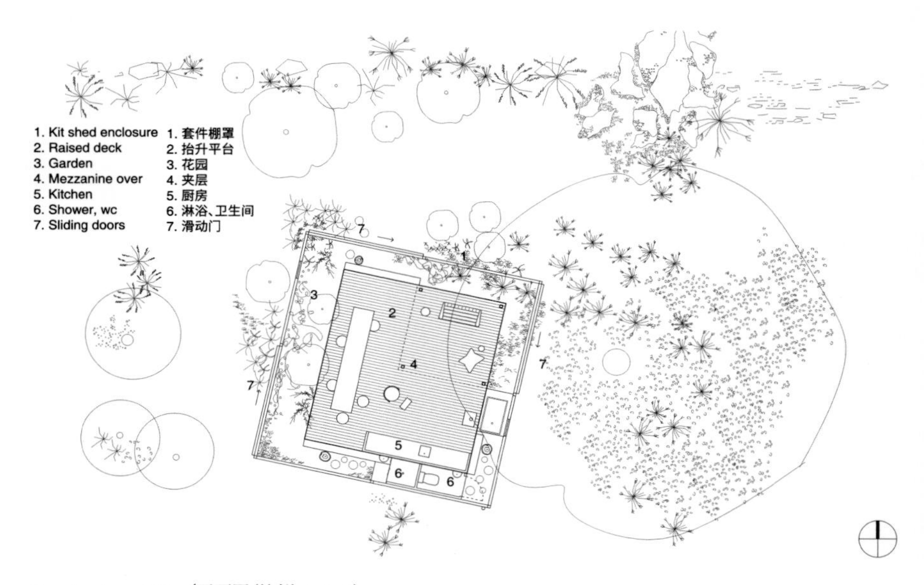

Plan (scale: 1/250)／平面图(比例:1/250)

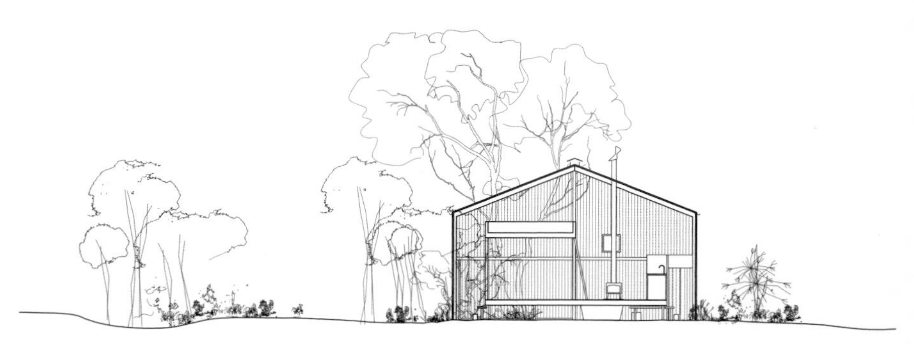

Section (scale: 1/250)／剖面图（比例：1/250）

Credits and Data

Project title: Garden House
Client: Louise Wright and Mauro Baracco
Location: Westernport, Victoria, Australia
Completion: 2015 (Structure), Landscape (ongoing)
Architect: Baracco and Wright Architects
Design Team: Louise Wright, Mauro Baracco, Catherine Horwill
Consultants: Perrett Simpson, Melbourne Sheds & Garages
Project area: 73 m²
Project estimate: $50,000 AUD

With this small building the architects aimed to explore a semipermanent structure that was conceived through the inclusion of the house in the site's ecosystem.

The site is part of a leftover heavily vegetated corridor in between cleared grazing land. A historical anomaly, it gives a glimpse into what used to be there, although it too now is mostly altered through domestic gardens, human and (non native) animal activity. Small patches of the endemic vegetation remain, mainly tea tree heath, among mown grass, introduced species and plants considered invasive weeds. Half of the street is trying to support these patches through seed collection and dispersal, weeding and connecting. The other half of the street routinely chop down trees and mow down any emergent endemic species that might pop up in their lawn.

The site connects to its neighbouring vegetation and Westernport – a large tidal bay. The road now occupies the position of an ephemeral creek, and being downhill, the area can be seasonally wet and dry, and can flood. It acts as a compromised wildlife corridor for animals travelling from nearby Gurdies and Grantville Nature Conservation Reserves to the coast, and perhaps most successfully supports birds.

On this site the existence of endemic terrestrial orchids indicated that the soil had not been altered, and that, embedded in the soil and under the introduced grass were the bones of a plant community that once grew there. With an aim to support the strength of the remnant indigenous vegetation present on the site, weeding was carried out using the Bradley method – which works from 'good' outward, so that slowly the vegetation can re-establish.[1] What occurred is a new type of balance. Isolated from the overall web of relationships and systems, it is difficult to know if you are strengthening a plant community or if it will now always depend on you. However, as mentioned elsewhere in this catalogue/book, the presence of Greenhood Orchids (Pterostylis Nutans) are evidence of the presence of Mycorrhiza (required at the germination stage) and often a symbiotic relationship with certain trees. Mycorrhiza fungi is a crucial foundation for healthy soils and have recently been credited with the network used by trees to communicate with each other.[2] Continued mowing and fertilisation of the (non native) grass on the site when it was bought, a hasty positioning of the house on their location, potential changes to the hydrology of the site such as water penetration/overland flow/microclimate, and symbiotic tree removal among others would have meant these orchids would have disappeared

pp. 44–45: The holiday house sits on a strip of remnant vegetation among cleared farmland surroundings in a coastal area. All images on pp. 44–53 by Erieta Attali unless otherwise noted.

第 44-45 页：这座度假屋坐落于沿海地区被整齐农田包围着的、一块狭长的残余植被带上。

in the near future.

By observing and supporting the expansion of small remnant patches of endemic vegetation, a shape of the site emerged that revealed an area where no regeneration was occurring, and as it turned out had been the site of imported fill, effectively smothering the seed stock and altering the soil.

The house was situated on this ‘clearing’, raised above the ground to allow for overland flow and so as not to ‘cut’ the site. Apart from a small utilities area, no ground is sealed.This supports the expansion of the vegetation inside the house. Now part of this ecosystem, this house supports life. The disturbance generated by the construction was quite minimal,[3] but nonetheless enough to generate the expansion of tea trees (which respond to disturbance). Tea trees now regularly grow inside.

The decision to make a building that admitted a lot of light, was in order to sustain plant life. Extremes are controlled through summer use of shadecloth, and also the strategic planting of the endemic trees that will in time further shade the house.As a holiday, and experimental house, it is conceived as just a little more than a tent: a deck and raised platform, covered by a transparent ‘shed’, the interior perimeter ‘verandah’ is garden space and living areas are dynamic yet subtly spatially defined; up, down, under, above. The soil and natural ground line are maintained and carried through.

Notes

1. See Australian Association of Bush Regenerators (AABR) n.d., Working with natural processes: the Bradley method, Australian Association of Bush Regenerators, viewed 10 February 2018, http://www.aabr.org.au/learn/whati-bush regeneration/generalprinciples/the-bradley-method.
2. See Wohlleben, P 2016, The Hidden Life of Trees: What They Feel, How They Communicate – Discoveries from a Secret World, Blanc Inc., Carlton, Victoria, Australia.
3. Through an unusual application of garage kit construction technology, very little disturbance or waste was created (around one cubic metre of waste, the majority of which was recyclable).

Opposite, above: The living deck is raised 800mm allowing floodwater to pass underneath, and the standalone mezzanine contains the bed. Opposite, below: The polycarbonate cladded, steel-framed house is kept to its minimal with only a raised deck, timber mezzanine, small fireplace and a bathroom.

对页，上：居住平台高出地面 800 毫米，洪涝排水可以从下面通过，床具放置于独立的夹层；对页，下：钢结构房屋由 PC 塑料（聚碳酸酯）覆盖，仅包含架高的平台、木质夹层、小壁炉和浴室等最简化要素。

通过这座小型住宅，建筑师旨在探索一种能够将房屋融入场地生态系统的半永久性结构。

在人工打理的放牧场之间有遗留一块狭长地带，这里被繁茂的植被覆盖着，项目场地就处于其中。这一处非常规的历史遗留地带，能够让我们窥探到过去的景象，尽管这里已经受到家庭花园、人类和非本土动物活动的影响而几乎被彻底改变。在修剪过的草坪、引进树种和被认为是入侵者的杂草中，仍保留着小片的本土植被，主要是野生茶树。半条街的住户正试图通过收集种子、散播、除草并使这些小片植被连成群落，从而保护它们。剩下半条街的住户依旧照常砍伐树木，并割除在他们的草坪上新冒出来的本土植物。

该场地连接着附近的植被和西港——一个大潮汐湾。这条路现在占据着一条季节性小溪的位置，并且是下坡道，该地区也会出现季节性的潮湿和干燥，偶有发生洪水的可能。对于要从附近的格迪斯和格兰特维尔自然保护区迁徙到海岸的动物们来说，它不失为一条野生动物走廊，同时也可能为鸟类提供栖息地。

在这片场地上，本地特有的陆生兰花的存在表明土壤并没有发生改变，而且，在引种的草丛下面的土壤中，埋藏着曾经生长在这里的植物群落遗存。为了增强场地如今仅存的本土植被的生长力，这里采用布拉德利法进行除草——从“生长良好的区域”逐渐向外作用，以便慢慢地恢复这些植被[1]。而后，这里达到了一种新的平衡。独立于整个关系网和系统之外，你很难知道自己是在强化一个植物群落，还是使它往后永远依赖于你。然而，正如书中其他地方提到的那样，存在绿帽兰就证明存在菌根（萌芽期所必需的），也证明它们常与一些树木有着共生关系。菌根真菌是健康土壤的重要基础，最近也被认为是树木之间用来相互沟通的网络[2]。如果继续给场地上那些买来的非原生草种修剪施肥、潦草地决定建房子的位置，都将给诸如渗水、地表水流、微气候等场地水文环境带来潜在变化。除此之外，如果在众多树木之中，被移除的偏偏是与菌根共生的那棵，那可能意味着这些兰花在不久的将来将不复存在。

通过观察和帮助残余的小片本地植被扩张，场地上显现出一片未出现植物再生的区域，事实证明这里是引入的外来植物种植区，它们有效地抑制了种子的储备并改变了土壤。

房子正是建在这片“空地”上；同时，房屋高于地面使地表水可以流通，以免“截断”场地与外界的连接。除了一小块设备区域外，其余地面都没有进行防水处理，这使得植物在室内也能生长扩张。现在，这所房子已成为这里生态系统的一部分，为生命提供支持。施工产生的干扰非常小[3]，但仍然足以引起茶树的扩张（这是对干扰作出的反应）。现在，茶树正常地生长在室内。

决定建造一座能吸收大量光的建筑是为了维持植物的生长。夏季极端天气是通过使用遮阳布来调节的。此外，通过策略性地种植地方特有树种，随着时间流逝，树木长大，也能更好地为房子遮荫。作为一个度假屋和体验式房屋，它的构思不止于一顶帐篷：一个架高的木质平台被一个透明的“棚”罩着，围绕着室内的“游廊”成为花园；起居空间既充满活力，又在空间上有着巧妙的界定；上方、下方、低处、高处交织。土壤和自然地平线得以保留并会一直延续下去。

注释

1. 澳大利亚灌木再生协会（AABR）n.d., Working with natural processes: the Bradley method，澳大利亚灌木再生协会，查阅于 2018 年 2 月 10 日：http://www.aabr.org.au/learn/whati-bush-regeneration/generalprinciples/the-bradley-method.
2. Wohlleben P, The Hidden Life of Trees: What They Feel, How They Communicate – Discoveries from a Secret World, Blanc Inc.，澳大利亚维多利亚州卡尔顿，2016。
3. 通过使用与众不同的车库套件建造技术，几乎没有产生场地干扰或废料（大约 1 立方米的废料，其中大部分是可回收的）。

pp. 50–51: The almost transparent wall blurs the boundary between the interior and exterior. The ongoing repair of the indigenous vegetation would eventually grow and provide shade and privacy for the house. This page: Slidable polycarbonate doors open up to bring the external environment within. Image by Lisa Atkinson.

第 50-51 页：几乎透明的墙使室内和室外的边界变得模糊。正在被修复的本土植被终将长大，并为房屋遮荫和提升私密性。本页：可滑动的 PC 塑料门将内部空间打开，同时将外部环境引入室内。

Crowd Productions

Flexi-frame House
Melbourne, Victoria 2007

群作工作室
灵活构架住宅
维多利亚州，墨尔本 2007

Credits and Data
Project title: Flexi Frame House
Client: Crowd Productions
Design Team: Michael Trudgeon, Anthony Kitchener, Costa Gabriel, Veronica Saunders, Gabriel Saunders, Karen Storey (Model)
Consultants: Richard Fooks (Structural engineer), VFV, Colder Products

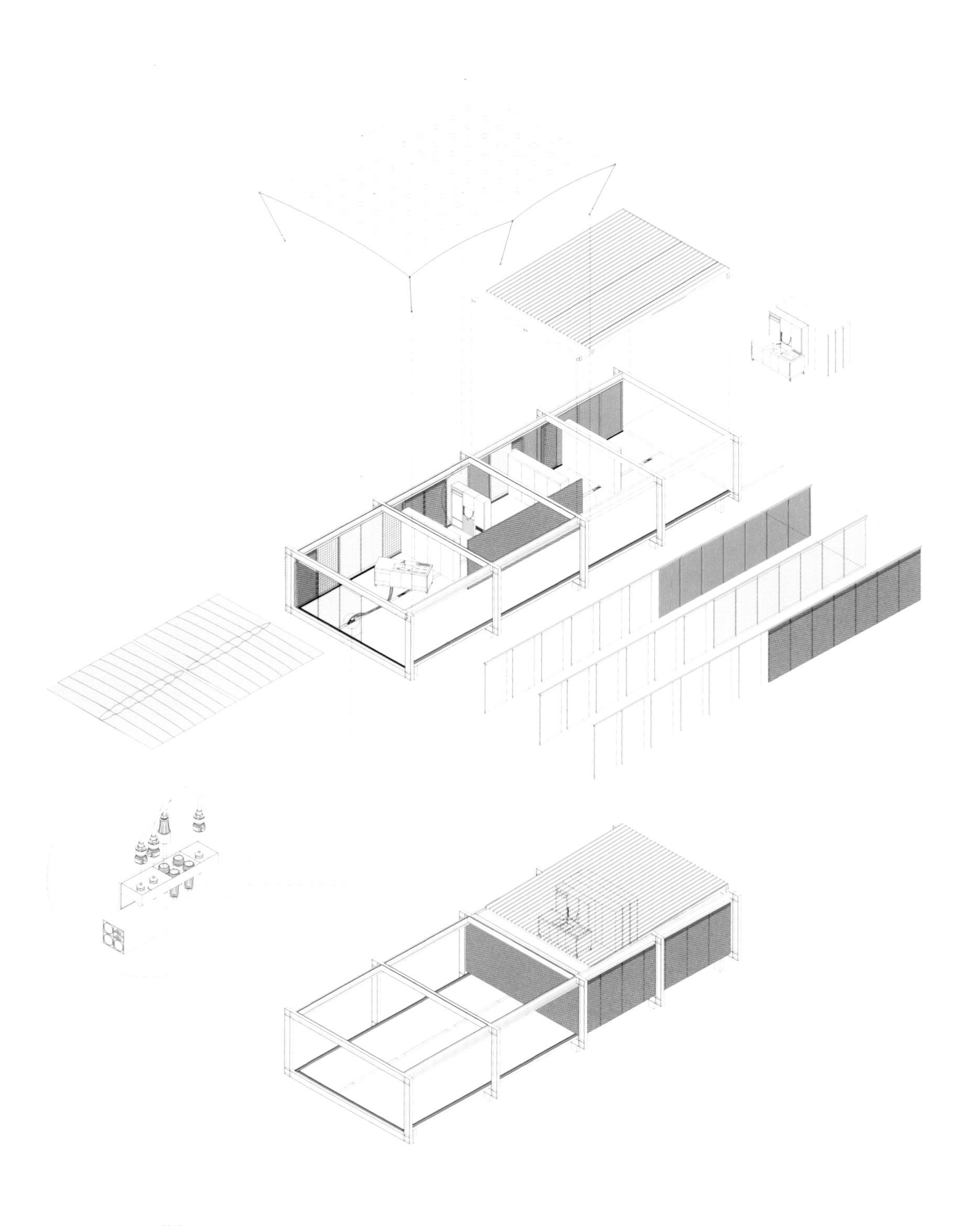

Frame diagram／构架图

Flexi-Frame House – is a compact living environment inspired by traditional rural Japanese domestic architectural planning. It comprises two pavilions: one for living and entertaining, and a more private zone for sleeping and storage. Conceived as a radically flexible and energy-efficient system, it claims very little space, uses minimal construction materials and offers multiple spatial configurations. All external walls are sliding panels, which can be packed away in the storage pavilion. The skin can be opened up to the landscape or closed down, providing protection from adverse weather.

A single integrated service loom, or channel, contains all the necessary services and runs orthogonally below the habitable aspaces. The loom conducts mains water and electricity in, grey water and sewage out, and supports telephone and data. Within a floor channel, it allows for flexible connection for any service via the universal plug-and-play connection system. When not in use, the house packs down into a storage shed and an outdoor deck with pergola.

Cirrus MVR Nomadic Water Conserving Bathroom – is a relocatable modular bathroom built around a central water recycling and purifying unit. The stainless steel module is equipped with a built-in bath, shower, toilet and hand basin, and provides all the services for contemporary bathing and ablution using a fraction of the water that current bathroom fittings require. An internal "weather cycle" continuously produces fresh water from the grey water output of the bath, shower and hand basin. This cycle replicates the sun's sterilizing rays, and the clouds' distilling of rainwater. The project uniquely combines chemical, medical and industrial food processing technologies for application to a domestic-scale integrated bathroom appliance.

Mobile Hyper Kitchen Technology Demonstrator – is a box of cooking tools, around which people can gather and prepare food. It contains an induction cooktop, an oven, a filtered downforce air extractor, a dishwasher, a microgreens growing rack, up to three sinks, two chopping blocks, a water heater and two garbage bins, all packaged up into a compact, mobile pod. This self-contained kitchen on wheels can move to wherever it is needed, and is operable wherever electricity and cold water are available, and if possible, a drain or sink within 40-m to remove the grey water from the sink and dishwasher. The design deploys and integrates interdisciplinary technological innovations ranging from scientific processes to space station, aircraft, yacht and car design.

Both the Cirrus MVR and Mobile Hyper Kitchen are flexible servicing infrastructure modules. They employ universal connectors to access water and sewage inlets, outlets, data and power, almost anywhere in a building. Fixed wet service zones such as kitchens and bathrooms are thus recast as movable.

p 54: Physical model of Flexi-frame house. Opposite, above: Prototype of the Mobile Hyper Kitchen Technology Demonstrator. Opposite, below: Render of the Cirrus MVR Nomadic Water Conserving Bathroom.

第 54 页：灵活构架住宅实体模型。对页，上：超级移动厨房技术演示器原型；对页，下：Cirrus MVR 可移动式节水浴室效果图。

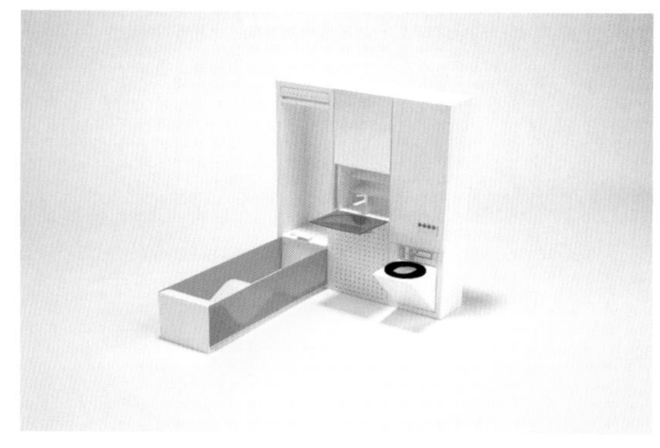

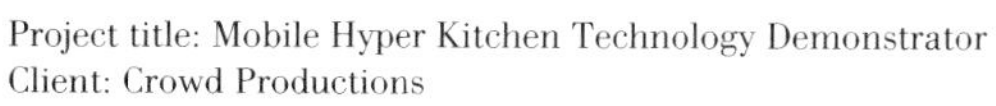

Project title: Mobile Hyper Kitchen Technology Demonstrator
Client: Crowd Productions
Design Team: Michael Trudgeon, Michael Loreto, Tracy Tung, Richard Benson, Joe Loreto, Anthony Kitchener, Jonno Swan, Adrian Loreto, Hanyi Lin, Costa Gabriel, Glynis Nott, Veronica Saunders, David Poulton, Joseph Brabet, John Burne, Roger Halley, Adelle Lin
Consultants: VFV, Colder Products, Rainbow Filters

Project title: Cirrus MVR Nomadic Water Conserving Bathroom
Client: Crowd Productions
Design Team: Michael Trudgeon, Anthony Kitchener, Costa Gabriel, Veronica Saunders, John Burne, Vaughan Howard, Joseph Brabet (Model)
Consultants: VFV, Colder Products, Rainbow Filters
Axonometric drawing (Mobile Hyper Kitchen)
Axonometric drawing (Cirrus MVR Nomadic Water Conserving Bathroom)

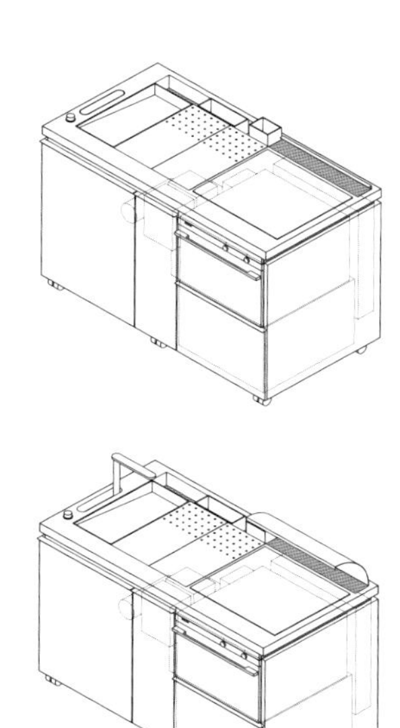

Axonometric drawing (Mobile Hyper Kitchen)
轴测图（超级移动厨房）

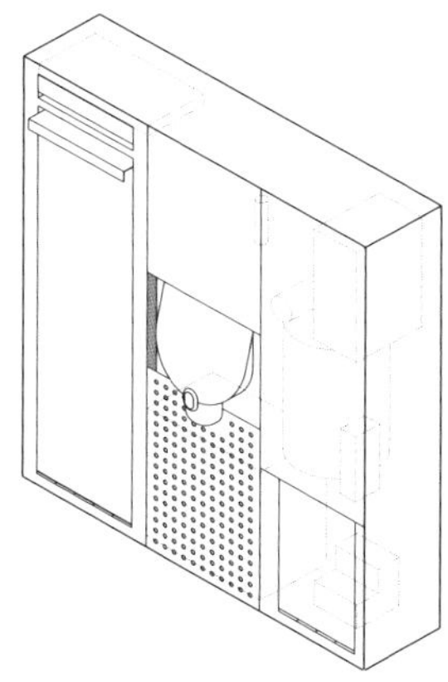

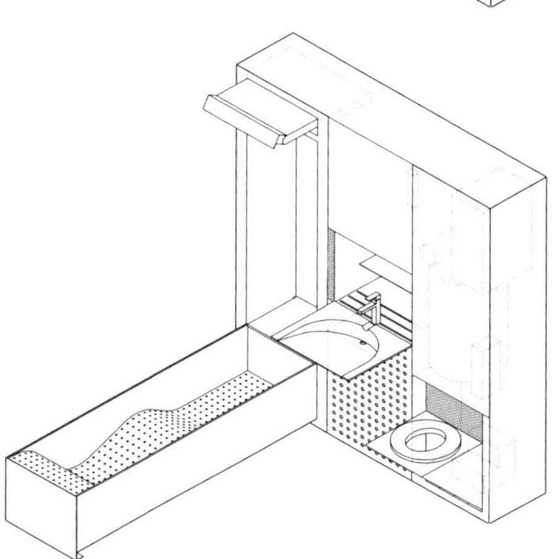

Axonometric drawing (Cirrus MVR Nomadic Water Conserving Bathroom)
轴测图（Cirrus MVR 可移动式节水浴室）

D01, D02, D03 SLIDING ALONG CONTINUOUS TRACK

DO1, DO2, DO3滑动门
沿连续轨道

WT1

GR1

GR1

FF1

MS3

SH1

SH1

MS3

SH1

MS3

D02 SLIDING ALONG CONTINUOUS TRACK

D02滑动门
沿连续轨道

WB1 ABOV

屋面板1 上

D04

GR1

WT1

D01D02, D03 SLIDING ALONG CONTINUOUS TRACK

DO1, DO2, DO3滑动门
沿连续轨道

Floor plan (scale: 1/100)／平面图(比例:1/100)

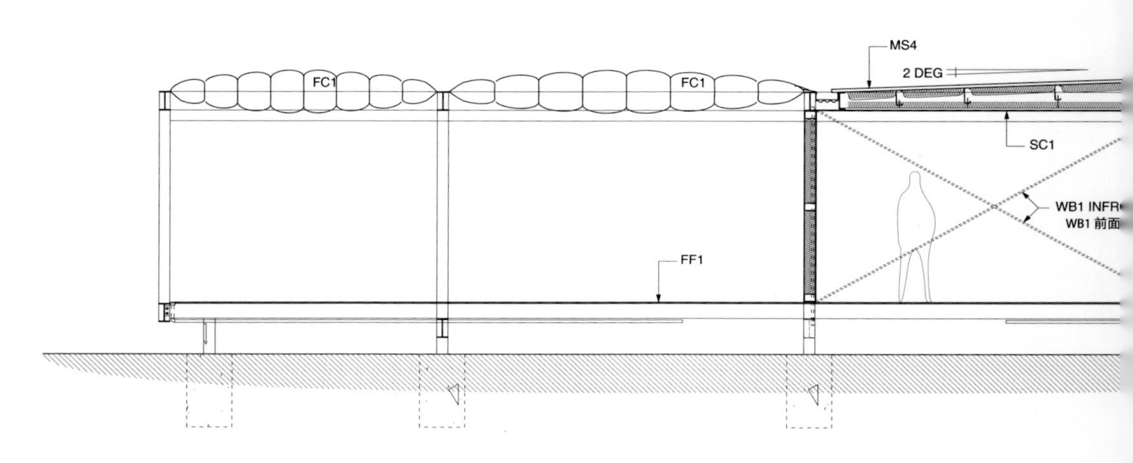

Long section (scale: 1/100)／纵向剖面图(比例:1/100)

灵活构架住宅——一种紧凑型居住环境，灵感来自日本传统的乡村住宅。它包含两个亭式建筑：一个用于起居和娱乐，另一个作为更私密的空间，用于睡觉和储物。它被认为是一种非常灵活且节能的系统，占用的空间很小，使用的建筑材料可做到极小限，并且可提供多种空间格局。所有外墙都是可滑动的嵌板，且可以通过滑动被收纳至储藏间内。外墙既可以对周围的景观敞开，又可以在恶劣天气中通过围合，为室内提供保护。

一个包含了所有必要系统的集成式服务机，或称服务带，在居住空间下方正交运行。服务机输入自来水和电力，排出灰水和污水，并提供通信和网络支持。仅通过地面的预设服务带，就可以用万向插头灵活地连接任何设备，接入即可使用。当不使用时，住宅可以被缩减至一个简易储物棚和一个带凉棚的室外平台。

Cirrus MVR 可移动式节水浴室——一个可迁移的模块化浴室，围绕中央水循环净化单元而建。不锈钢模块配有内置浴缸、淋浴、坐便器和洗手池，可满足现代沐浴清洁的一切需求，且相较于市面上的浴室设施，用水量极少。内部的“气候循环”不断地将浴缸、淋浴间和洗手池排出的灰水转化为淡水。这个循环复制了太阳的杀菌光线，以及云层的雨水蒸馏。该项目将化学、医学和工业食品加工技术独特地结合在一起，实现了一个家用规模的集成浴室装置。

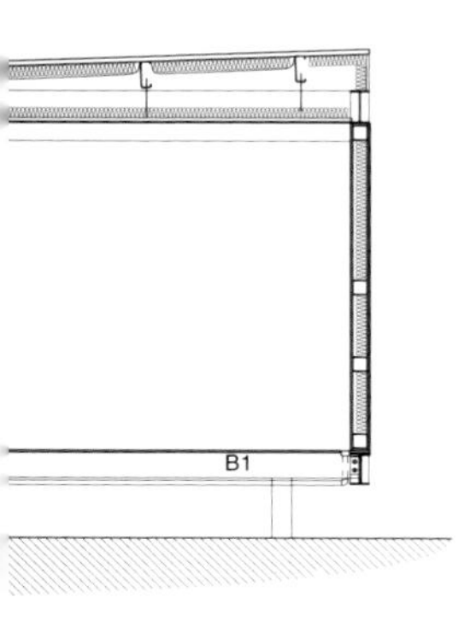

超级移动厨房技术演示器——一个人们可以在此聚集并准备食物的厢式烹饪工具集成体。它包含一个电磁炉灶、一个烤箱、一个可过滤的下吸式油烟机、一个洗碗机、一个微型蔬菜种植架、不超过三个的水槽、两个砧板、一个热水器和两个垃圾桶，以上所有都包含在一个紧凑的移动舱内。这个带轮子的独立厨房可以移动到任何需要的地方，并且可以在任何有水电供应的地方使用，如果 40 米范围内有排水口或盥洗池，还可以排掉水槽和洗碗机中的生活灰水。从科学方法到空间站、飞机、游艇和汽车设计，该设计有效地利用和整合了跨学科的技术创新。

Cirrus MVR 和超级移动厨房都是灵活的基础服务设施模块。他们使用通用的连接器，几乎能在建筑的任何位置接入供水、排水、网络和电力。由此，厨房和浴室这类固定的湿服务区被重新设计成可移动的模块。

Troppo Architects

Robe – trop_pods

Robe, South Australia 2017

特罗普建筑师事务所

罗布特罗普分离舱

南澳大利亚州，罗布 2017

Credits and Data

Project title: Robe – trop_pods

Client: Private

Location: Robe, South Australia

Design: 2016

Completion: 2017

Architect: Troppo Architects

Consultants: PTD Design (Structural), tecon (Certifier)

Builder: Oscar Building

Fitout & Furniture: Oscar Furniture

Project area: 126 m^2

Project estimate: $220,000 AUD

Trop_pods are a creation of Troppo Architects, a nationally and internationally acclaimed practice and winner of the 2010 Global Sustainable Architecture Award. All trop_pods come with troppo's high-level, environmentally responsible service, to fit them to a particular site and always-unique client needs.

Trop_pods are a system of portable and flat-pack spaces that come together to create thoughtful, tough and durable architecture. Pods work for residential and tourism purposes, both as stand-alone independent elements or in aggregation as more expansive buildings or complexes. They can also deliver small offices and meeting rooms, as characterful and environmentally connecting workplaces.

The creation of trop_pods spans five years of skill sharing collaboration with Oscar Building. Located in regional Victoria in southeastern Australia, on Australia's highway network, Oscar are third generation furniture makers and builders, with a strong ethos of innovation and community service.

Prefabrication of the pods enables high-level trade quality and material efficiency. Oscar's systems include for waste minimization and recycling; and their facility is solar powered to offset all production electricity demands. Prefabrication also significantly reduces site environmental impacts by comparison with a traditional build over many months.

On site, each pod is lifted onto a concrete-free footing system; pop-outs are extended; hinged decks lowered; verandahs raised; and eaves and awnings are 'clipped on'. (And of course all can be unclipped, lifted and removed to their next future.)

There is a lot to be said for taking with you only what you can carry, and trop_pods are much the same. Building to what can fit on the back of a single truck requires us to consider every square meter while delivering clever, usable spaces. Yet, whilst small, they're made generous by their broad connection to place and setting.

Robe is located on a coastal rural property in South Australia's South East, three quite separate pods strew along a vegetated dune site to yield living and bedroom pavilions and a "lookout" bedroom. They afford guest accommodation alongside an existing dwelling, and capture the informality of a beach holiday, surrounded by space and nature. Each pod is carefully planned and founded to fit with site levels and to orient for solar control and views, and to establish privacy between each.

pp. 60–61: General view of two of the three trop_pods located on a rural coastal site in Robe. Opposite: View from the Living Pod verandah. All images on pp. 58–65 by courtesy of the Architect.

第60-61页：三个特罗普分离舱坐落在罗布乡村沿海地区，图为其中两个分离舱的整体外观。对页：从起居舱游廊看到的景色。

Site plan (scale: 1/150)／总平面图（比例：1/150）

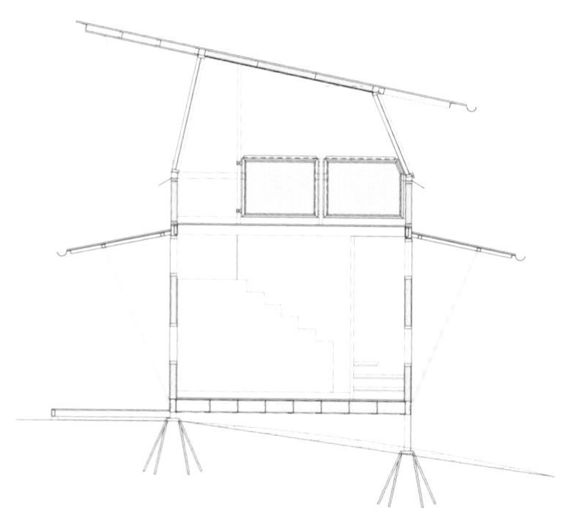

Lookout pod section (scale: 1/120) ／观景舱剖面图（比例：1/120）

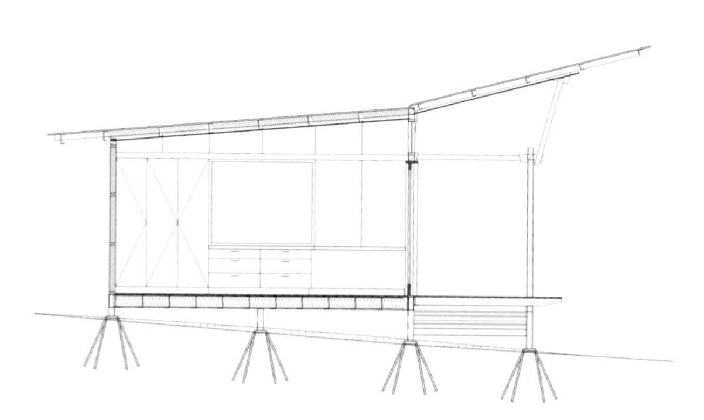

Bedroom pod section ／卧室剖面图

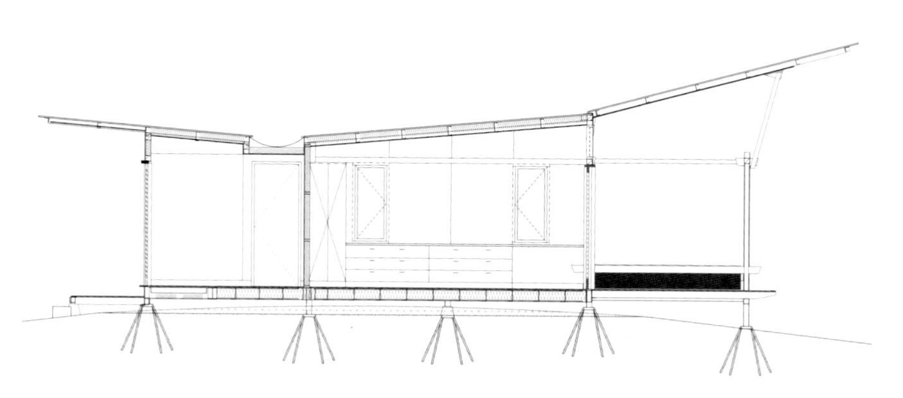

Living pod section ／起居室剖面图

特罗普分离舱是特罗普建筑师事务所的作品，该事务所是一家享誉国内外的建筑事务所，也是“2010 年全球可持续建筑奖”的得主。每一个特罗普分离舱都配有特罗普高品质且对环境负责的服务，以此适用于特殊的场地和客户个性化的需求。

特罗普分离舱是一种便携、平板式包装的空间组合系统，这些空间聚集在一起形成周到、坚固而耐久的建筑。舱体适用于住宅和旅游，既可以作为结构独立的单体装置，也可以作为集合体形成更宽敞的建筑或综合体。它们还可以用作小型办公室和会议室，成为有特色且与环境连通的工作场所。

特罗普分离舱的创造背后是与奥斯卡建筑长达五年的技术共享与合作。奥斯卡建筑位于澳大利亚东南部的维多利亚州，地处澳大利亚高速公路网之中，是第三代家具制造商和建造商，具有很强的创新精神和社区服务精神。

舱体的预制可实现较高的贸易质量和材料效率。奥斯卡采用的方法包括废物最少化和回收利用；他们的设施采用太阳能供电，以抵消所有制造用电需求。与持续多月的传统建造相比，预制还可以显著减少对现场环境的影响。

在现场，每个舱体都被悬吊放置在无混凝土的基脚结构上，伸出的部分被展开，铰接的木质平台被降低，游廊被抬高，屋檐和遮阳棚便捷地附上去。当然，所有东西都可以被拿下来，抬起并移至下一个地点。

“轻装上阵”这句话包含了很多学问，这同样适用于特罗普分离舱。要建造出可以放进一辆卡车里的建筑，我们需要仔细思考每一平方米的使用，从而创建一个巧妙、便于使用的空间。它们尽管体积很小，但由于与场地和环境之间的丰富联系而变得十分宽敞。

罗布位于南澳大利亚州东南部沿海的乡村地区，三个非常独立的舱体沿着一座植被覆盖的沙丘散布开来，形成起居室、卧室，以及一个“观景”卧室。它们在现有住宅旁边提供客用住宿，四下开阔，充满自然气息，也饱含轻松随意的海边度假氛围。每个舱体都经过精心的设计和建造，以适合现场地平、日照和景观视野，并保证个体之间的私密性。

p. 64: View of from the bedroom of the Lookout Pod. Opposite: View from the kitchen of the Living Pod.

第64页：从“观景”卧室看到的景色。对页：从起居室厨房望向室外。

Troppo Architects

Anbinik – trop_pods

Kakadu National Park. Northern Territory 2015

特罗普建筑师事务所

安比尼克特罗普分离舱

北领地，卡卡杜国家公园 2015

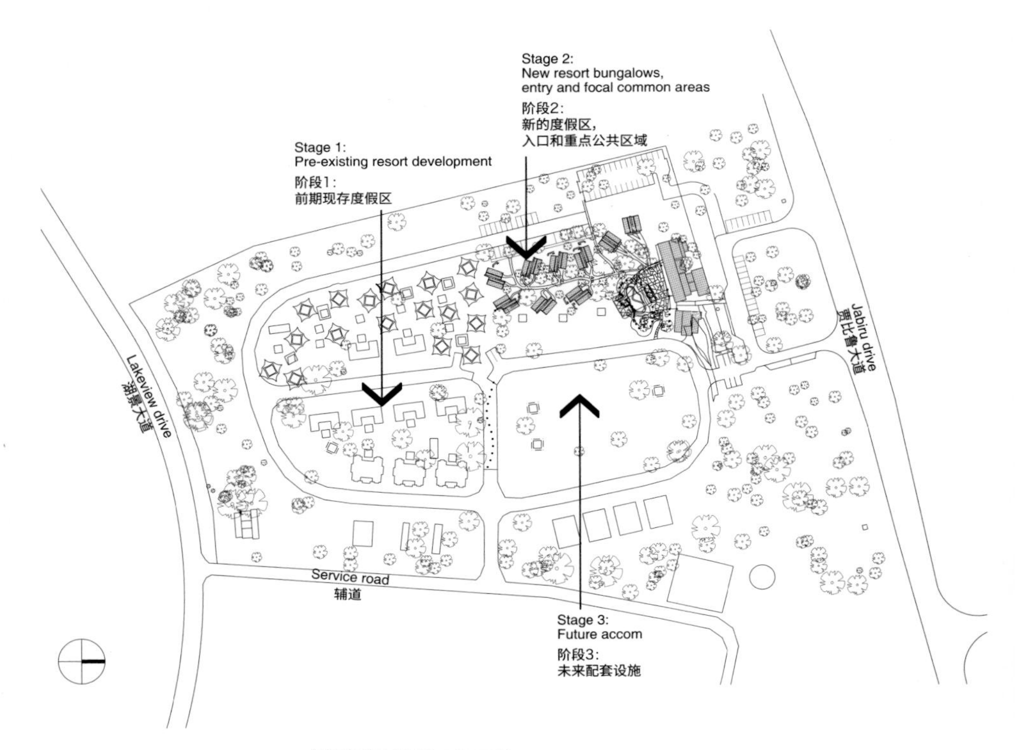

Overall site plan (scale: 1/3,000)／总平面图（比例：1/3,000）

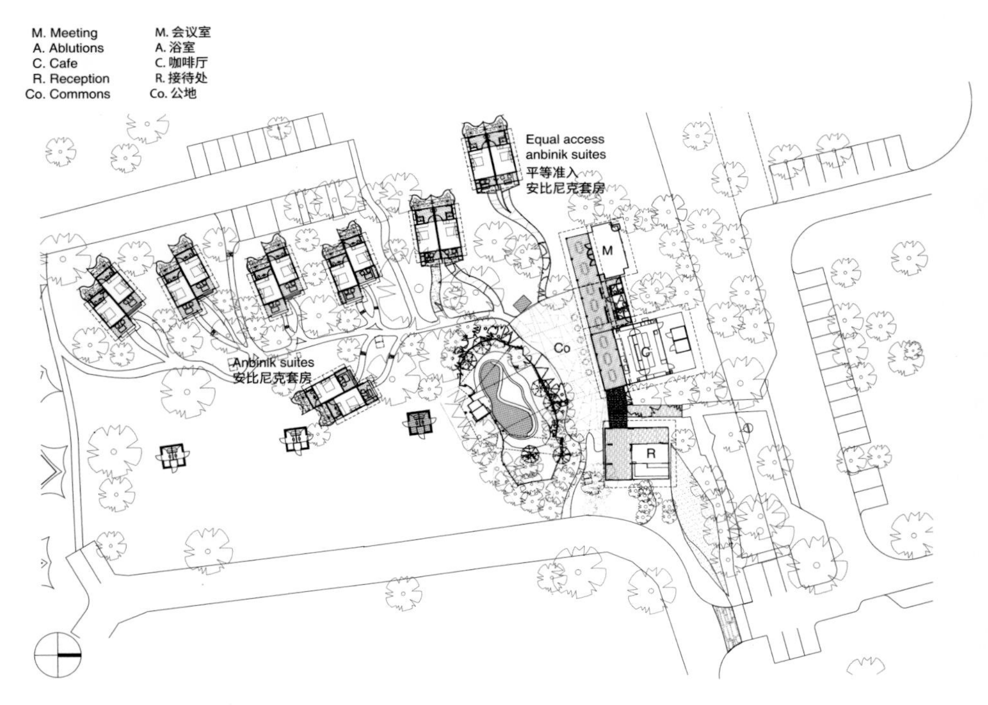

Site plan (scale: 1/500)／基地总平面图（比例：1/500）

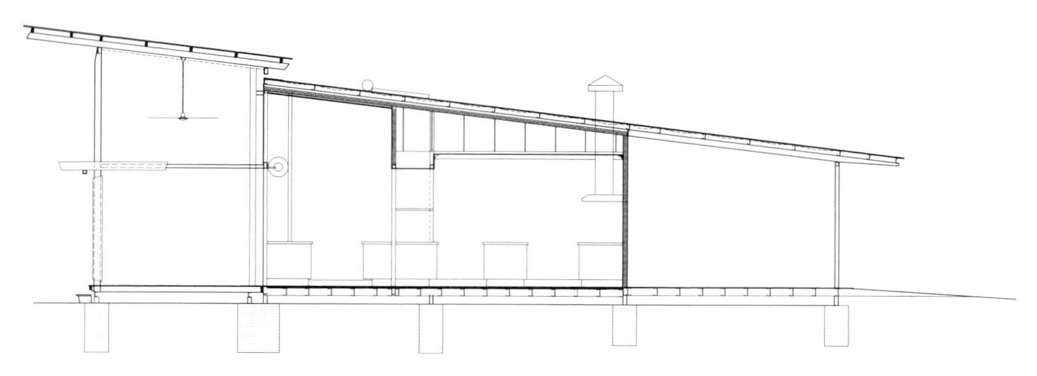

Commons and meeting section (scale: 1/150)／公共空间和会客空间剖面图(比例:1/150)

pp. 68–69: View of the trop_pod units located along a pathway in a tropical garden setting in Jabiru. This page: Common space with a cafe and meeting area for its resort guests. All images on pp. 68–75 by courtesy of the Architect.

第68-69页：特罗普分离舱组合住宅的外观，该建筑位于贾比鲁热带花园内的一条小径旁。本页：公共空间内设有咖啡厅和会客区，供度假村的客人使用。

The precursory trop_pods project was 14 accommodation units for Anbinik Kakadu Resort operated by Kakadu's Djabulukgu Association in Australia's Northern Territory. Troppo have worked with Djabulukgu since 1993.

Like the great spreading An-binik (Allosyncarpia) of Kakadu's secret gorges, the resort offers a unique shelter experience. A cool, shady retreat from the heat of the Top End day. These "tropical bungalows", together with a cafe, reception and meeting areas, completed in 2014, represent a final stage of development of the resort.

The bungalows were constructed as portables and shipped to the site. They are plasterboard free and completed with local art and bespoke furnishings, including pandanus inspired "kakadu" fretwork ceilings and Japanesey "noren" split curtain privacy blinds. We designed the chairs, stools, and tables, again, affordably.

All materials are "real", earthy and robust. Bathrooms open broadly to private landscaped courtyards and verandahs are grand, with daybeds. The bungalows are designed for deep shade and cross ventilation. The pool includes a raw steel "escarpment" backdrop and retractable reed shades.

From Anbinik, troppo developed a suite of pod types that can combine for diverse purposes, with each pod also being readily adaptable to play its site-fitting role, like in Robe (See pp. 60–67). Other basic precepts are for the pods to be built of "real" materials of Australian sourcing, to age gracefully, and with space-saving and nifty fitout, down to bespoke furniture.

Credits and Data

Project title: Anbinik – trop_pods
Client: Djabulukga Association
Location: Lakeside Drive, Jabiru Northern Territory, Australia
Design: 2011
Completion: 2015
Architect: Troppo Architects
Consultants: WS Consultants, R.E Proud (Structural), WGA (Civil Engineers), Peter Horne/ Alec Gangur (Services Engineer), QS services (Quantity Surveyor), tecon (Certifier)
Builder: Oscar Building, Aldebaran Contracting, Kakadu Contracting, Blueridge
Fitout & Furniture: Oscar Furniture, KM Custom
Project area: 12,500 m^2
Project estimate: $3.2 million AUD

Opposite, above: Exterior view of the trop_pod duplex unit. Opposite, below: Interior view of the unit's bedroom.

对页，上：连栋式的两座特罗普分离舱住宅外观；对页，下：单元卧室内部。

特罗普分离舱项目的前身是位于澳大利亚北领地安比尼克卡卡杜度假村的 14 个住宿单元，这里由卡卡杜地区的加布鲁古联合会运营。自 1993 年以来，特罗普建筑师事务所便开始了与加布鲁古联合会的合作。

就像仅在卡卡杜的隐秘峡谷中生长蔓延的安比尼克树（也称松节油树，是澳大利亚北领地特有的一种雨林树种）一样，该度假村也提供独特的住宿体验。结束了一天在炎热的北领地北端地区的游览后，这里将是一处凉爽、遮荫的避暑胜地。这些“热带小屋”与咖啡厅、接待区和会客区共同于 2014 年竣工，代表着度假村开发的最后阶段。

这些小屋被预制为易装卸的形式，并运输到现场。它们没有石膏板墙，而是靠当地艺术和定制家具共同建造，其中包括灵感源自露兜树的“卡卡杜”镂空天花板，以及由日式分体暖帘演化而来的私密百叶窗。我们也设计了经济适用的椅子、凳子和桌子。

所有材料都很“实在”、质朴和坚固。 浴室可通向私人园景庭院，游廊宽敞并设有躺椅。小屋配备了很宽的遮阳棚并且可对流通风。游泳池包括一面粗钢打造的、如“悬崖”一般的幕墙和可收起的芦苇遮阳棚。

特罗普建筑师事务以安比尼克分离舱为基础，开发出了一套可以组合用于多种用途的分离舱样式，每个舱体也能够很容易地发挥其场地适应性，像在罗布那样（见第 60-67 页）。舱体还拥有其他基本特征：舱体及定制家具均由澳大利亚本土的真材实料建造而成，随着时间的推移而越发美丽，既节省空间又精美实用。

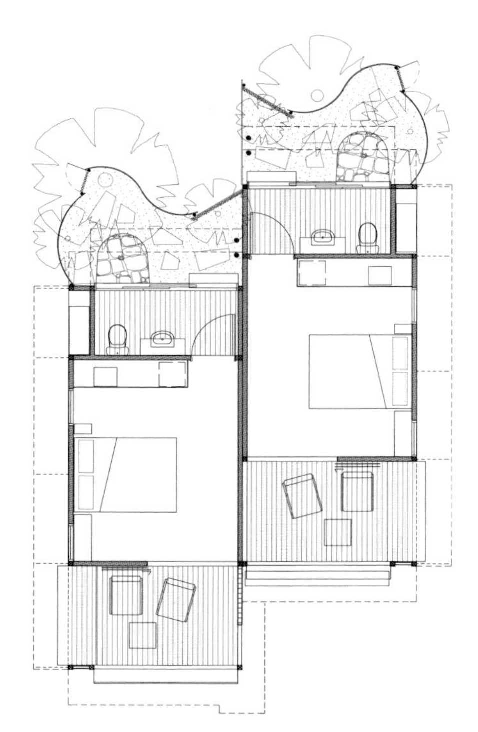

Typical bungalows plan (scale: 1/150)／小屋平面图(比例:1/150)

Opposite, both images: Prefabricated trop_pod units are transported to Kakadu National Park before being reassembled at its final location.

对页，两张：预制的特罗普分离舱在最终位置组装之前，先被运送到卡卡杜国家公园。

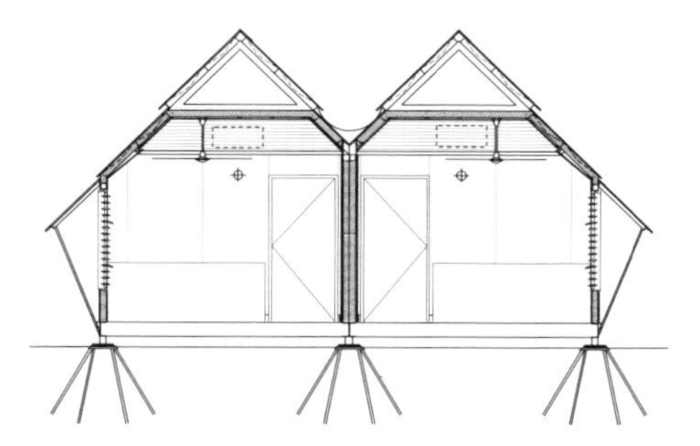

Bungalows short section／小屋横向剖面图

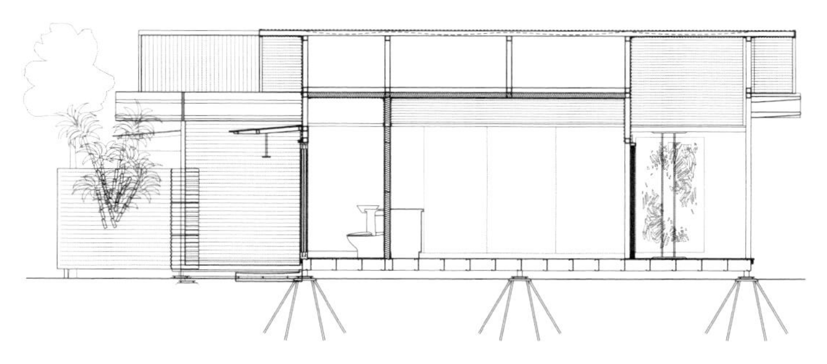

Bungalows long section (scale: 1/150)／小屋纵向剖面图(比例:1/150)

Andresen O'Gorman Archit

Mooloomba House

North Stradbroke Island, Queensland 1998

安德烈森 · 奥格曼建筑师事务所

穆鲁姆巴住宅

昆士兰州，北斯特拉德布罗克岛 1998

Credits and Data

Project title: Mooloomba House

Client: Brit Andresen and Peter O'Gorman

Location: Point Lookout on Minjerribah (North Stradbroke Island), Queensland, Australia

Design: 1996

Completion: 1998

Architects: Brit Andresen and Peter O'Gorman

Consultants: Graham Mellor (Carpenter), John Batterham (Engineer), Lon Murphy (Contractor)

Project area: 585 m²

Mooloomba House is a two-storey timber house built at Point Lookout on Minjerribah, also known as North Stradbroke Island, within the subtropical climatic region of South East Queensland. Located on a hillside, the site is flanked by groves of banksias and folding topography.

The architecture amplifies the qualities of the site and explores the expressive capacity of construction through metaphor, geometry and the material properties. The house design both defers to the existing landscape and alludes to a mythical or idealised landscape in which to live. Much of its 500-squaremetre floor area is conceived as outdoor rooms, loosely flanked on three sides by a cloister walkway. The open-weave lines of the timber structure and its gray-green colouring recall the immediate forest and suggest a continuous landscape.

Systems of proportion, structural interdependence and aesthetic balance are explored relative to the specific properties of local hardwoods. Eucalyptus has a great capacity for durability and strength, but equally, hardwood will warp, twist, cup and crack as it dries after milling, resulting in its more common use – as framing and behind cladding. A simple strategy is adopted to tame the excessive lateral movements through the vertical lamination of thin members of opposing grain formation, and the integration of a 1,200mm-wide wall panel of 18mm waterproof ply sandwiched in between. The frame acts as enlarged cover battens to the joints in sheets. A generative 2,000mm × 1,200mm (1:1.618) proportion of the expressed panel and primary frame achieves visual integrity.

In the benign climate of coastal Queensland, the hardwood assembly system can be designed for transparency. This opportunity is pursued through integration of technology (material and technique), light, territoriality (the plan) in relation to the landscape. Varied construction systems define diverse spatial experiences; from light bower-like spaces, to protective, cave-like enclosures. The extraordinarily rich environmental context allows the architecture to mediate between the real and the idealised.

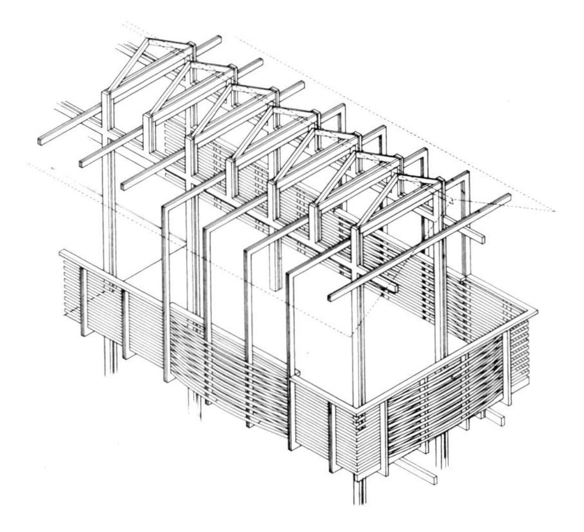

Structural axonometric drawing／结构轴测图

pp. 76–77: Exterior view looking towards the perched sleeping alcoves on the upper floor and sand courtyard on the lower floor. The grey-green timber structure and its open-weave lines reflects its forest settings. Opposite, above: The belvedere on the upper level is accessed by an external timber staircase. The intricacy of the timber work found here evocates this fragility that matches itself with its environment. All images on pp. 76–83 by Michael Wee unless otherwise noted. Opposite, below: View of the walkway located independently on the east. Image by Reiner Blunck.

第 76-77 页：位于上层的壁龛式卧室和下层沙地庭院的外观。灰绿色的木质结构和交错裸露的线条体现出其所处的森林环境。对页，上：上层的观景台可通过外部木质楼梯进入。仅在此处才能找到的精细木质工艺唤起了一种与环境相称的精致感。对页，下：独立于东面的人行通道。

Opposite: View of the double-height living room. The roof made from corrugated acrylic enhances the sense of transparency. This page: Interior view of the perched sleeping alcove along a narrow walkway on the upper floor. p. 83 : The small and contained belvedere looks out onto the shoreline and ocean.

对页：两层通高的客厅，由波纹亚克力制成的屋顶增强了通透感。本页：在上层狭窄的走道看壁龛式卧室的内部。 第 83 页：紧凑而隐蔽的观景台可眺望海岸线和大海。

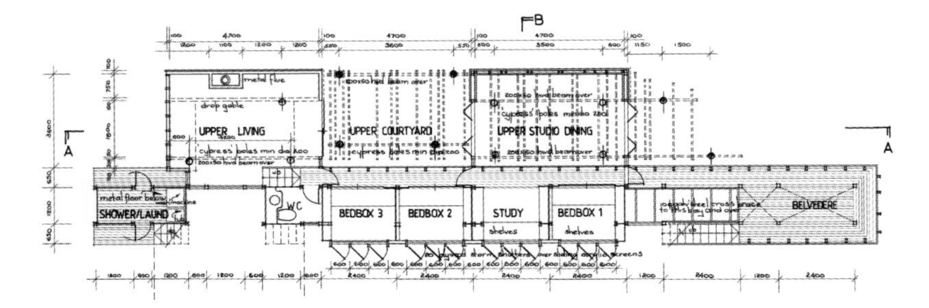

Upper floor plan／上层平面图

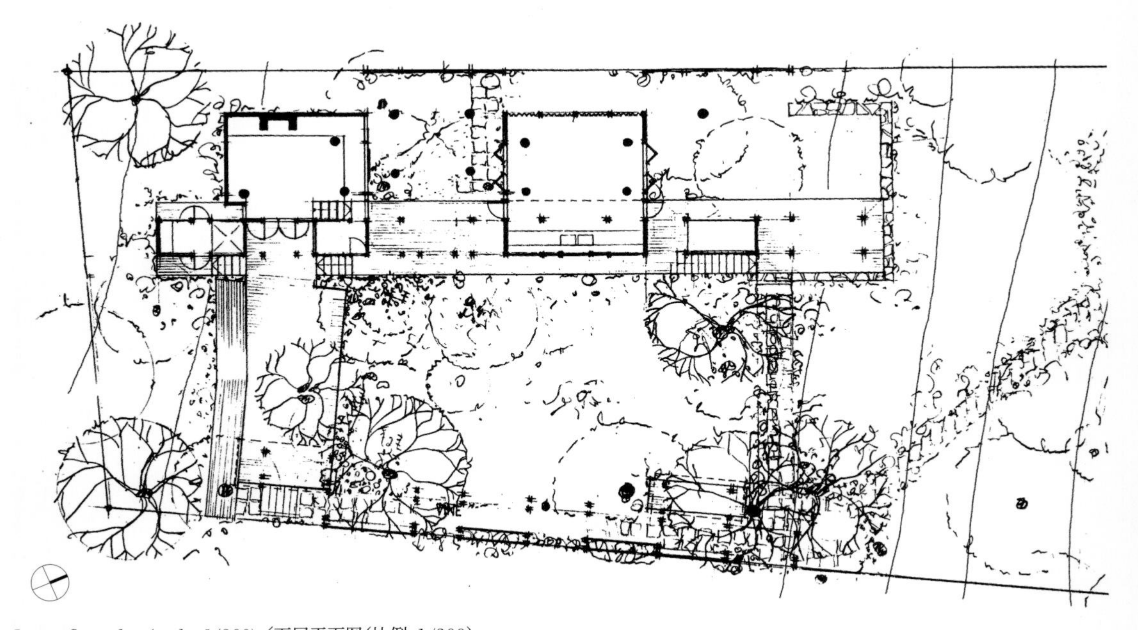

Lower floor plan (scale: 1/300)／下层平面图(比例:1/300)

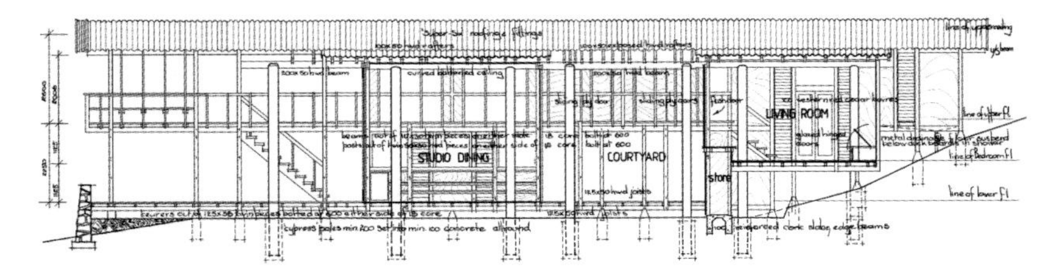

Section A (scale: 1/300)／A剖面图(比例:1/300)

穆鲁姆巴住宅是一座两层木结构住宅，建于明杰里巴（Minjerribah，土著语），也称为北斯特拉德布罗克岛的岬角观景点上，位于昆士兰州东南部的亚热带气候区。该场地处于山坡上，两侧是山龙眼（澳大利亚一种常绿灌木）树丛和褶皱地形。

这座建筑放大了场地的特性，并通过隐喻、几何形状和材料特性探索建造中的表现力。住宅的设计既尊重现有景观，又影射出一种神话或者说理想化的景观，以供人栖居其中。其 500 平方米的建筑面积大部分被设计成室外空间，三面都松散地设有回廊步道。交错裸露的木结构及其灰绿的颜色让人联想到周围的森林，并由此感知一种连续的景观。

基于本地硬木的具体特性，建筑师对比例、结构的相互依存系统和美学平衡进行了探究。桉树具有极好的耐久性和强度，但同时，经过铣削的硬木在干燥过程中会发生翘曲、扭曲、凹陷和开裂，这也使其更常被用作框架和不外露的结构。过度的横向位移可通过简单处理来控制，即将纹路构造相对的一组薄板进行垂直压制，两组压制薄板之间夹着一块宽 1,200 毫米、厚 18 毫米的防水层板。框架充当了放大版的封边压条，覆盖着薄板间的连接。上述面板和主框架形成了 2,000 毫米 ×1,200 毫米 (1:1.618) 的比例，实现了视觉的完整性。

在昆士兰州沿海的温和气候下，硬木装配系统可以通过设计被赋予通透性。建筑师把握住这次设计机会，将与景观技术（材料和技法）、光线和景观地域性规划融合起来。不同的建造方法定义了多元的空间体验；从轻质的凉亭般的空间，到具有保护性、洞穴般的内部空间。极为丰富的自然环境使建筑可以在现实和理想之间得到统一。

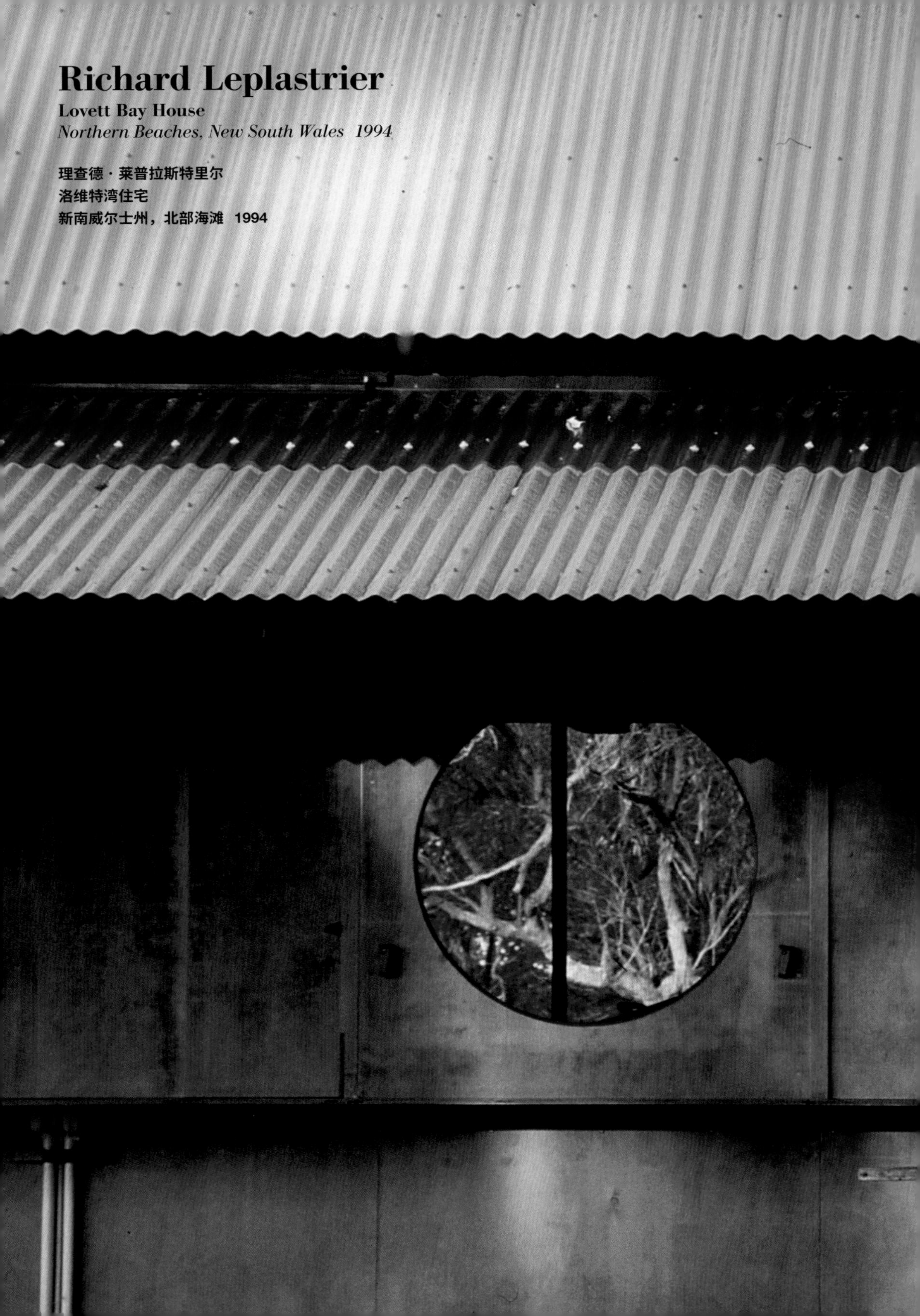

Richard Leplastrier

Lovett Bay House

Northern Beaches, New South Wales 1994

理查德·莱普拉斯特里尔

洛维特湾住宅

新南威尔士州，北部海滩 1994

This small house is located in Lovett Bay, on the fringe of Sydney's Ku-ring-gai Chase National Park along the western shores of Pittwater, a sunken river valley. Although on the mainland, access is only by water. The local climate is tempered by the Pacific Ocean and the warm East Coast current, which mediates the mild subtropical nature. The site receives power and is otherwise autonomous.

The project explores architecture as a vessel and a cradle for the family. The form of the building is elemental, conceived as a single room within a "greater room", the walls of which are the surrounding cliffs, and the floor the shifting tidal level of the river below, which rises and falls nearly two meters every six hours.

The house was handmade by the architect and master craftsman Jeffrey Broadfield, together with friends, and evokes their mutual engagement with ancient joinery methods and boat building technology. Lack of vehicular access to the site informed a lightweight and easily assembled timber construction, designed to be fully dismantlable for future relocation and reuse. The operable perimeter walls are multilayered, to allow the room to transform in degrees of enclosure. Together with the generous extension of the roof over a column-free verandah, they provide both protection and unbroken connectivity to the natural surroundings. The system allows the passive adjustment of the house to variations in natural light and prevailing breezes, the seasons and inclement weather.

A timber bath and a modest kitchen are accommodated on the verandah, and frame pleasurable rituals of everyday life within the natural setting. The inner room at once operates as a workroom, a sleeping area, living space and dry store. Embodied in this elemental house for Karen, Richard and their family is the search for an essential architecture that is economical, yet expansive. The creative life of this architect and his family gains strength and intensity in this extraordinary house.

Credits and Data

Project title: Lovett Bay House
Client: Karen Lambert and Richard Leplastrier
Location: Lovett Bay, Pittwater, Sydney, Australia
Completion: 1994
Architect: Richard Leplastrier
Construction Team: Jeffery Broadfield, Lee Hillam and friends

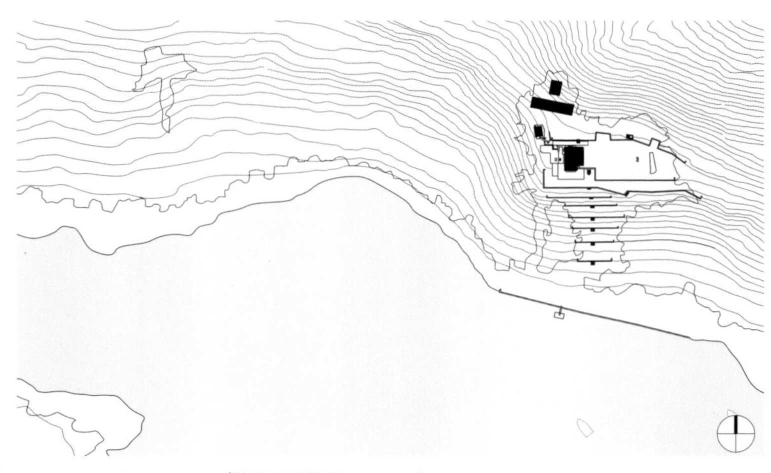

Site plan (scale: 1/3,000)／总平面图(比例:1/3,000)

pp. 84–85: Elevation view of the house with its windows framing the surrounding landscape. Opposite, above: The house overlooks into the river valley from the south. Opposite, below: View of the house from the east. All images on pp. 84–91 by Leigh Woolley.

第 84-85 页：住宅的立面视图，窗户定格了周围的景色。对页，上：这座住宅从南面俯瞰河谷；对页，下：从东面看到的住宅外观。

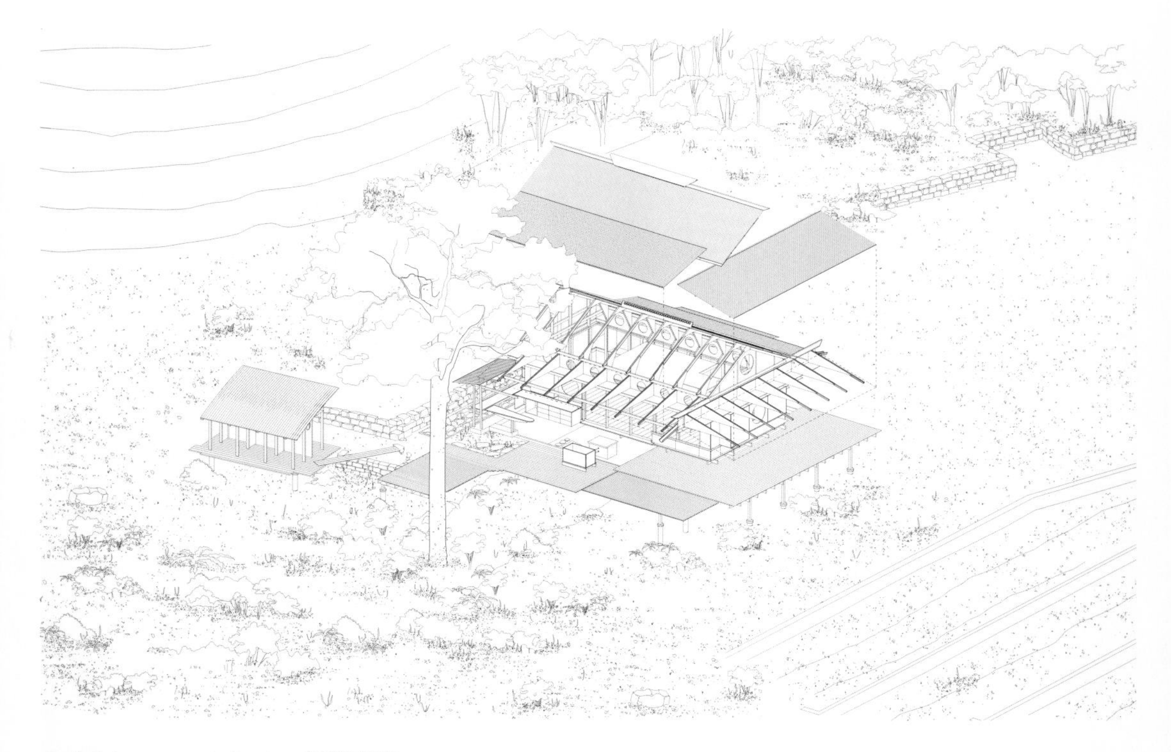

Exploded axonometric drawing／轴测展开图

这座小型住宅位于洛维特湾，地处悉尼库灵盖狩猎地国家公园边缘、凹陷的彼特沃特河谷西岸。尽管住宅在大陆上，却只能通过水路进入。当地气候受太平洋和温暖的东澳大利亚暖流影响，使其具有温和的亚热带气候特征。该场地仅需接受供电，其他方面皆为自主运营。

该项目的设计理念是将建筑视作摇篮般的家庭聚集地。其基本的形态被构想成“大房间”内的一个单独房间，大房间的墙壁是周围的悬崖，地面是下方河流每六小时涨落近 2 米的、不断变化的潮汐水位。

这座住宅是由建筑师和工匠大师杰弗里 · 布罗德菲尔德与朋友们一起动手建造的，这也促使他们与传统细木工技法和造船技术相借鉴。车辆无法进入现场，造就了这一轻质且易于组装的木质结构。该结构亦可被完全拆解，以便将来搬迁和再利用。四周的墙是多层的，可通过调节使房间的开放程度发生变化。无柱凉廊及延伸至其上方的宽阔屋顶，既提供了保护，又与周围的自然环境保持了不间断的连通性。这样的系统使房屋可以进行被动式调节，以适应自然光、盛行风和季节的变化，以及恶劣天气。

凉廊上容纳了木质的浴室和小巧的厨房，并在自然环境中营造出宜人的日常生活体验。内部空间可同时用作工作室、卧室、起居室和干燥的储物空间。在这座住宅为理查德、凯伦及其家人建造的小屋中，体现了对经济和宽敞的建筑本质的探寻。建筑师及其家人的创造性的生活可能会在这一强有力的居住空间中变得更加强大。

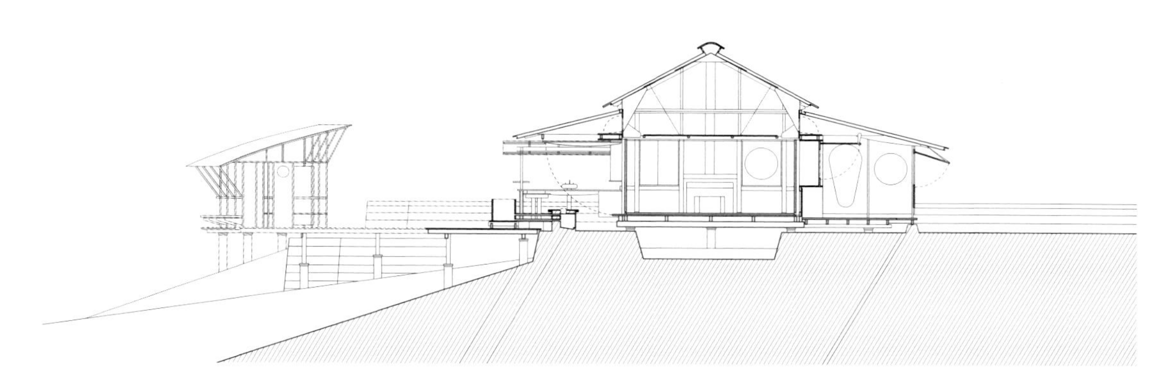

Short section／横向剖面图

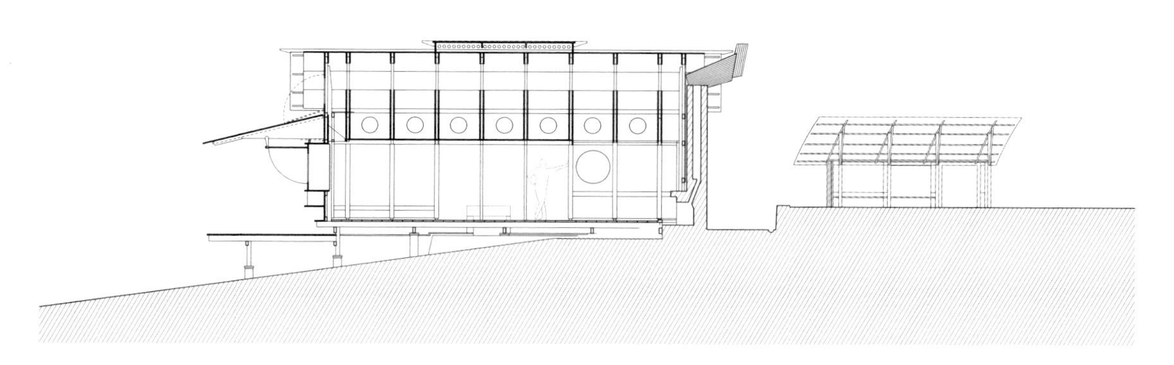

Long section (scale: 1/200)／纵向剖面图(比例:1/200)

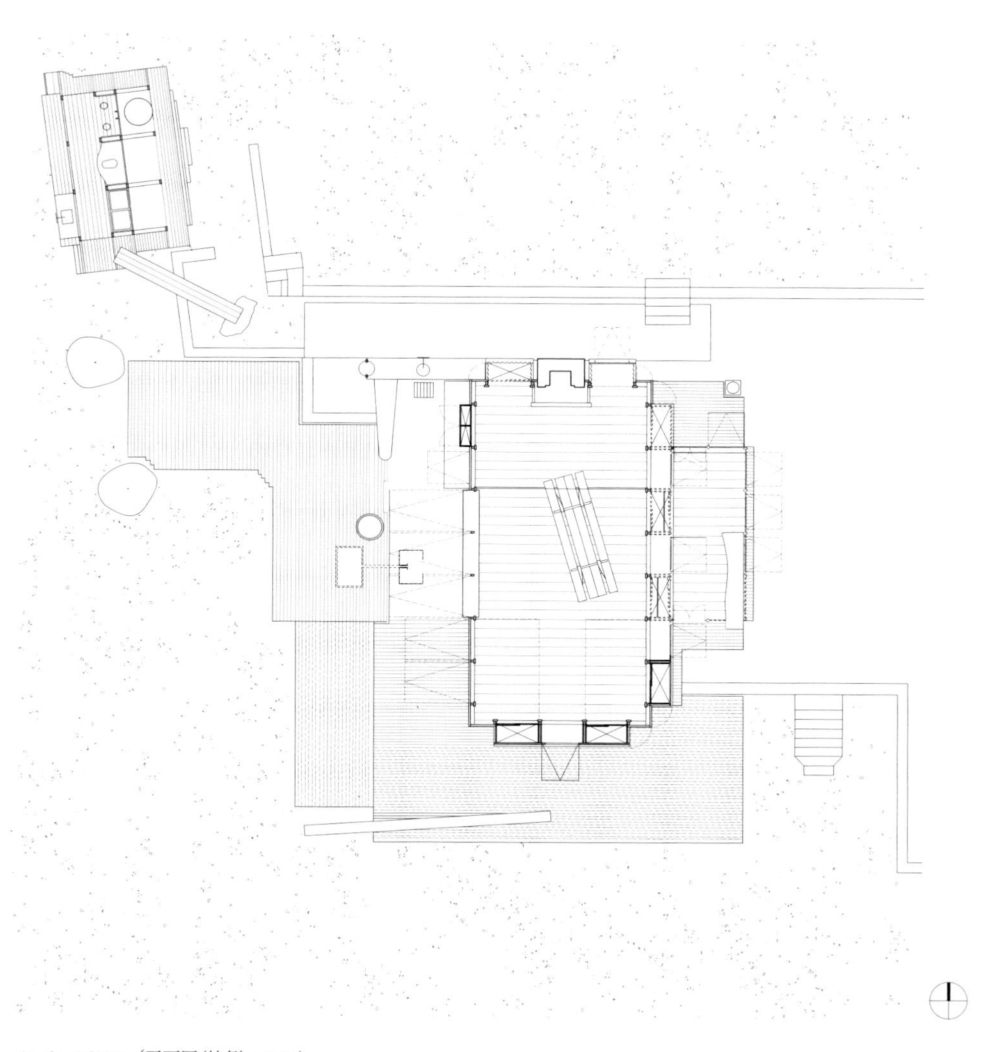

Floor plan (scale: 1/200)／平面图(比例:1/200)

Opposite, above: Living deck of the house. Opposite, below: Interior of the house with an open view of the landscape.

对页，上：住宅的起居平台；对页，下：拥有开阔景观视野的室内空间。

Glenn Murcutt

Donaldson House

Pittwater, New South Wales 2016

格伦·马库特

唐纳森之家

新南威尔士州，皮特沃特 2016

The house is located in Sydney's Northern Peninsula, Palm Beach. Almost hidden away from the road, the house sits on a steep slope in a native bushland within walking distance to the beach, opening up views towards both the land and sea. The upper level contains the entrance, foyer, main bedroom, bathroom, as well as the common spaces – kitchen, dining room and living room. The main bedroom is adjacent to the entry and orientated both north and to the rock to provide privacy from the street while taking advantage of the natural sunlight. A suspended concrete staircase with a large tapestry leads guests from the upper level down towards the street level. On the lower levels are three bedrooms, a laundry with a drying deck and two bathrooms set on three levels, having a more private ambiance nestled deep among native trees and bushland plants. Sited in a high fire risk zone, the house has to be designed according to strict regulations, from reinforced concrete foundations, toughened glass windows, steel roof purlins, sheet metal covered roof decking to walls and steelwork coated by dark grey micaceous paint. Despite such challenging conditions, this naturally ventilated house is sensitive to its environment with overhangs and operable windows providing shade and cross ventilation. It relies on solar panels covering the roof to provide its home's electrical needs and a water collection system recycles grey water throughout.

Credits and Data

Project title: Donaldson house
Location: Palm Beach, New South Wales, Australia
Completion: September 2016
Architect: Glenn Murcutt AO
Consultants: James Taylor & Associates (Engineer), Craig Poppleton (Builder)
Project area: 715 m²
Project estimate: $7.6 million AUD

pp. 92–93: View of the living room from the upper floor verandah. pp. 94–95: External details reflecting the architect's sensitivity to the landscape. Opposite: Concrete stairs leading down to main entrance of the upper floor. This page, above: View overlooking the roof of the house. This page, below: View of the upper floor verandah and the lower floor stairs that lead out into the nature. All images on pp. 92–107 by Anthony Browell.

第 92-93 页：从上层阳台看到的起居室。第 94-95 页：外部细节反映了建筑师对景观的敏感度。对页：混凝土楼梯向下通往上层空间的主入口。本页，上：俯瞰住宅的屋顶；本页，下：上层阳台和底层通向自然的楼梯外观。

这座住宅位于悉尼北部半岛的棕榈滩上，远离道路，坐落在一个被原生灌木覆盖的陡坡上。使用者步行即可到达海滩，可一览陆地和海洋美景。上层包含了入口、门厅、主卧、浴室以及公共空间，即厨房、餐厅和客厅。主卧与入口相邻，皆面朝北面岩石，这样既远离街道，保证了私密性，也能充分利用自然光线。一侧装饰有大挂毯的悬挑式混凝土楼梯，将客人从上层引导至下方的地面层。地面层包含了三间卧室、一间带晾晒平台的洗衣房和两间跨越三个高度的卫生间。空间若隐若现地处在原生树木和灌木植被之中，享有更加私密的氛围。由于该住宅位于火灾风险较高的区域，所以，从钢筋混凝土地基、钢化玻璃窗、钢屋顶桁条、金属薄板覆盖的屋顶平台，到均涂有深灰色云母漆的墙体和钢结构构件，都严格依据规范进行设计。即使在这样极具挑战的条件下，这座自然通风的住宅仍然保持了对环境的敏感度，采用飞檐及可开合的窗户来实现遮阳及对流通风，依靠覆盖屋顶的太阳能电池板来满足家庭的用电需求，并通过集水系统对所有灰水（译注：从洗脸盆和地漏里出来的水）进行回收利用。

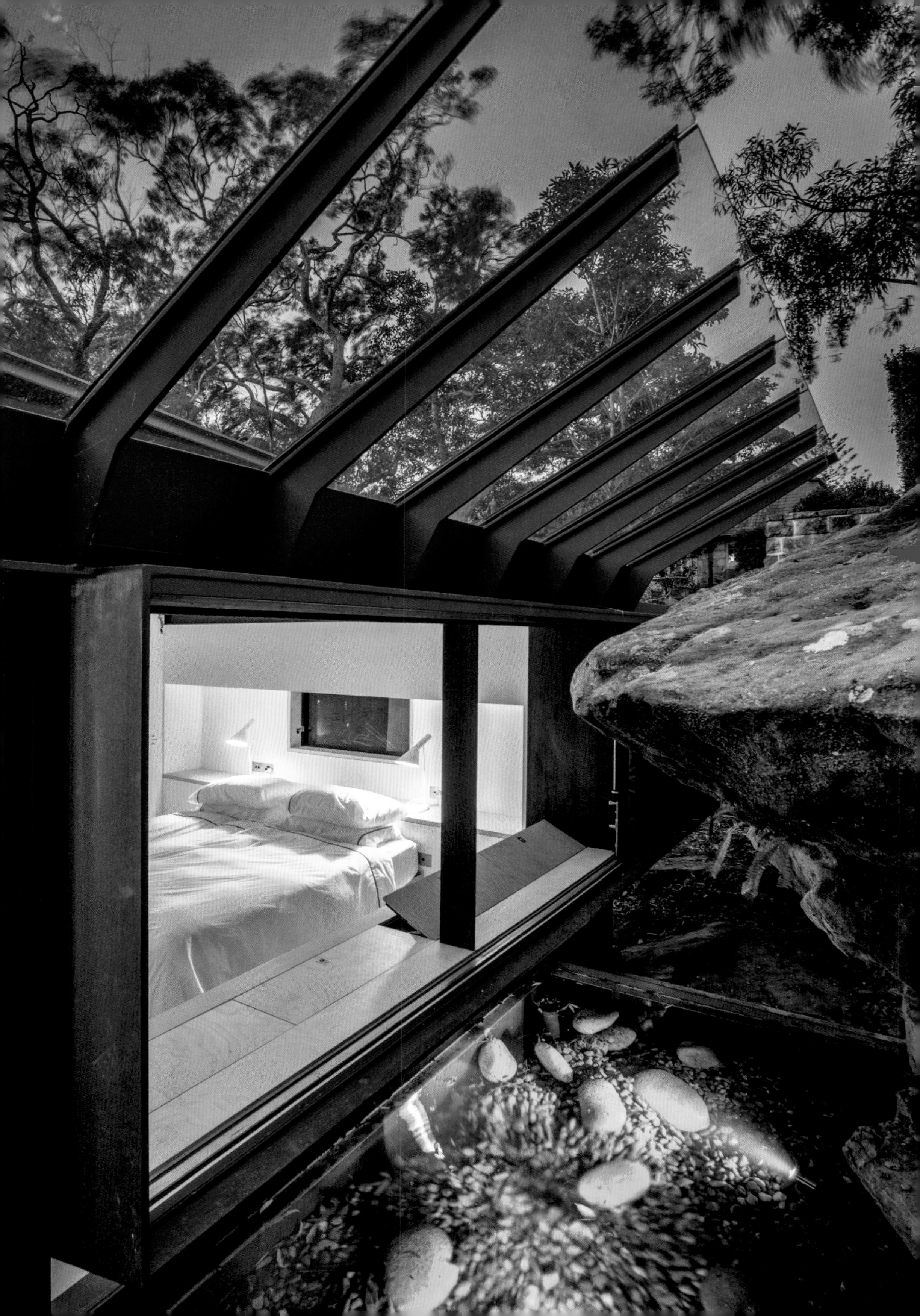

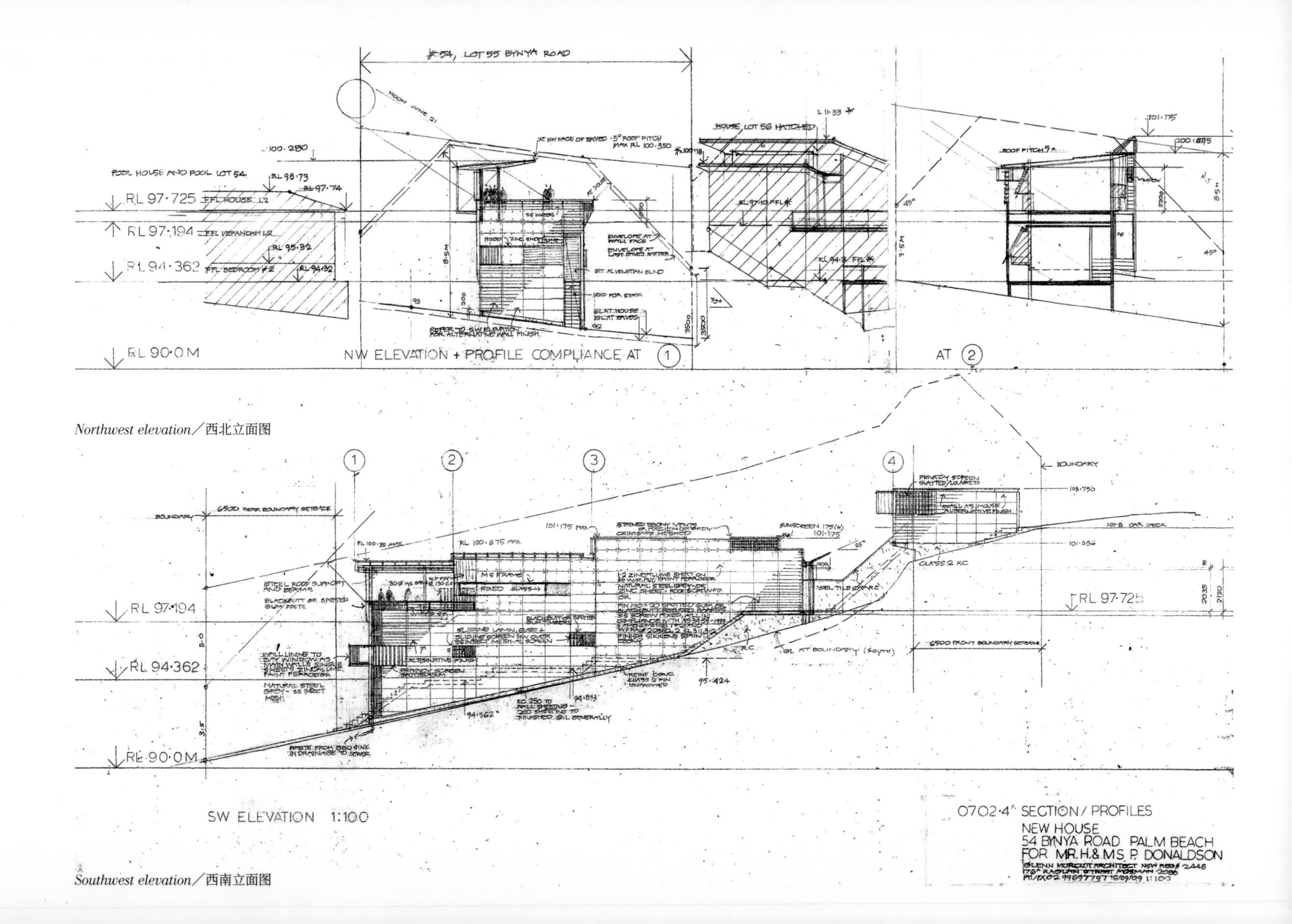

Northwest elevation／西北立面图

Southwest elevation／西南立面图

SOLAR HOTWATER PANEL

SUNSTOP GREY GLASS

EXTERNAL SUN CONTROL SCREENS OVER GLAZING

LAMINATED MILKY

ZINC ROOF FINISH

LAMINATED GLASS AND ALU LOUVRES

GLASS FIXED

PLANKS 100 WHITE SS FRICTION STAYS SPOTTED GUM/BLACKBUTT

VENTS

31-35 REC GAS BOOST FOR SOLAR HW

RWP

1·2 ZINCALUME SHEET ON 16 W/P PLY, PAINT FERRODOR NATURAL STEEL GREY - OR ZINC SHEET

RC

DECK

ZINC ROOF

EXTERNAL VENETIAN BLINDS, GREY ALUMINIUM

PIVOT VENTS CRIMSAFE BLACK SS MESH

BLACKBUTT TIMBER DOOR AND SCREENS, SPOTTED GUM DECK FOR CLOTHES RACK

PIERS RC - LOCATE TO AVOID TREE ROOTS

SHOWN ALTERNATIVE WALL FINISHES - DESCRIBED ON SW ELEVATION, DRG 0702·4

8·5

N.E. ELEVATION

1200

SIDE BOUNDARY

CLASS 2 RC WALL

SG SLATS (F) 40 X 20 @ 30 cc BIKKENS EBONY STAIN

ROAD AT BYNYA 'KERB'

SIDE BOUNDARY

1300

101·9

ADJ. SUNSCREENS OVER 6/4 OPAL LAMINATED GLASS

LEVEL OF HOUSE IN RELATION TO BYNYA RD

OUTRIGGER FOR BED 1 WINDOWS

97·725

WATER

POND 500 DEEP GAL MESH @ 290 RIVER GRAVEL @ 250 BELOW WATER LEVEL

SE ELEVATION TO BYNYA ROAD

Northeast elevation／东北立面图

8·5 HEIGHT PROFILE

35°

VENTS

RL 101·175

RL 97·725 FFL

2100

SHELF

MELAMINE

2101

45°

8·5M

95·75

CLOTHES DRYING DECK HW - SPOTTED GUM

95

RL 95·424 FFL

3·5

SECTION AA

RL 90·0M

3 PROFILE COMPLIANCE

103·7

RL 101·80

101·7

101

FIN 40 X 20 SPOTTED GUM/BLACKBUTT SLATS/POSTS @ 50 C.C. EBONY STAIN

1300

RC

103·75

CLASS 2 RC

99

104·61

103·75

101·47

101·086

CARPORT TO ADJACENT HOUSE LOT 56

4 PROFILE COMPLIANCE ENTRY SHR/STORE

0702·3A SECTION/PROFILES

NEW HOUSE

54 BYNYA ROAD PALM BEACH

FOR MR.H.& M.S.P. DONALDSON

GLENN MURCUTT ARCHITECT NSW REG# 2448

176A RAGLAN STREET MOSMAN 2088

PH/FX 02 99697797 28/09/09 1:100

Sections／剖面图

PLAN LEVEL 1 1:100

PLANTOUT WITH ADDITIONAL 20 ANGOPHORA COSTATAS DISTRIBUTED ON SITE AS SOIL DEPTHS ARE FOUND TO BE APPROPRIATE

LEVEL 1: RL 94·362 BED 2/BED 3
94·893 BED 4
95·424 LAUNDRY/BATH

CONTOURS & TREE LOCATIONS FROM SURVEY

0702·2 PLAN

NEW HOUSE
54 BYNYA ROAD PALM BEACH
FOR MR. H. & MS P. DONALDSON
GLENN MURCUTT ARCHITECT NSW REG# 2448
176A RAGLAN STREET MOSMAN 2088
PH/FX 02 99697797 15/09/09 1:100

Plan level 1／一层平面图

p. 98: General view of the house at night. p. 99: View from the master bedroom window looking towards the rockface and a pool of water protected by a tilted glass roof. p. 103: Main staircase of the house connecting all levels, from the living room on the top floor to the bedrooms on the lower floors. pp. 104–105: View looking towards the living room from the dining room. Opposite, above: Interior view of the master bedroom. With no air conditioning, monsoon windows (half opened in photo) provide natural ventilation even on a rainy day. Opposite, below: Interior view of bedroom on the lower floor, large angled windows on the north maximises daylighting and view. This page: View between the rock, looking towards the stairs leading down to the entrance on level 2.

第 98 页：住宅在夜晚的全景。第 99 页：透过有倾斜玻璃屋顶保护的主卧窗户可以看到岩石面和一汪池水。第 103 页：从顶层的起居室到底层的卧室，住宅的主楼梯连接着各楼层。第 104-105 页：从餐厅看到的起居室外观。对页，上：主卧内部。虽然没有空调，但季风窗（照片中季风窗半开）即使在雨天也能提供自然通风；对页，下：底层卧室内部，大角度倾斜的北面窗户将采光和视野最大化。本页：从岩石间望向通往下方二层入口的楼梯。

Wendy Lewin and Glenn Murcutt

Australian Opal Centre

Lightning Ridge, New South Wales 2015

温蒂·卢因与格伦·马库特

澳大利亚欧泊中心

新南威尔士州，闪电岭 2015

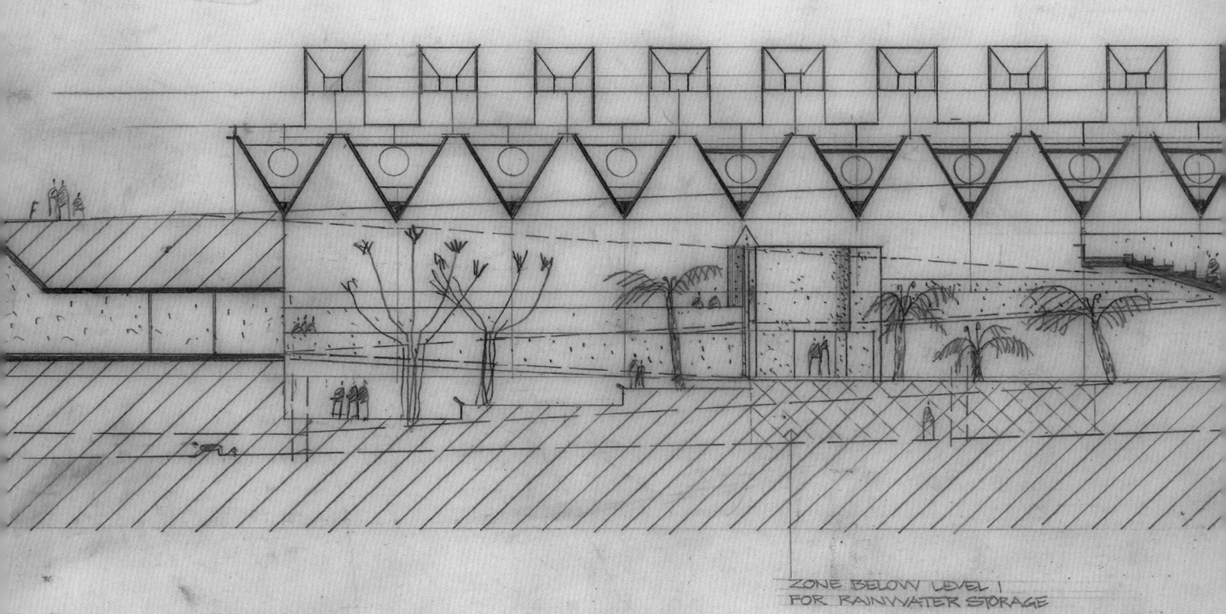

Credits and Data

Project title: Australian Opal Centre

Client: Lightning Ridge Opal Centre and the community of Lightning Ridge

Location: Lightning Ridge, central-northern of New South Wales, Australia

Design: 2015

Architects: Wendy Lewin and Glenn Murcutt

Project area: 31,000 m²

Project estimate: $34 million AUD

AUSTRALIAN OPAL CENTRE
LIGHTNING RIDGE

M
0 1 2 3 4 6 10

SECTION

ARCHITECTS: GLENN MURCUTT
WENDY LEWIN
ISSUE B
MAY 1, 2006
176A RAGLAN ST, MOSMAN. SYDNEY.

The proposed Australian Opal Centre (AOC) is a research and education facility for Lightning Ridge, an iconic mining town with the largest deposit of black opals in the world. Off the grid and without any town services, the site is adjacent to the historic 3-mile open-cut mine, the most extensively worked mine in the region. In this semi-arid region, summer temperatures can reach into a high of around 40 °C and winter temperatures can fall to below 5 °C.

Developed since 2003, with significant engagement in the intervening years by local government and the Lightning Ridge community, the AOC will support cultural and economic growth in one of Australia's most economically disadvantaged regions.The project has been conceived to be autonomous; it will generate its own power, support the collection, storage and recycling of water, the on-site management of waste systems, and incorporate passive heating and cooling. Substantially embedded, it takes advantage of the Earth's thermal mass, and the stable below-ground temperature of constant 22 °C to mitigate external heat loads on the building. Drawing on passive systems of cooling derived from ancient Middle Eastern desert architecture, malqafs draw air from the prevailing winds, down the chimneys and across a series of moist plates and bodies of water to provide cooler, conditioned air within the interior. Expressed externally, they assign the architecture a rhythmic monumental presence. Incorporating excavated material from the site into off-form concrete, the building will be "in and of" the site.

Within the interior, a gentle incline guides the visitor to a platform suspended within an open-cut mine, thereby revealing the ancient geological strata. The lower level holds subterranean exhibitions of Australian natural and cultural treasures, including the most significant public collection of rare Australian opals and opalized fossils globally. Displays extend to mining machinery, and the social history of Lightning Ridge. A garden supporting ancient plant species from the Gondwanan period features plants with a long fossil history, such as Wollemi pines, ferns and cycads.

Through the direct and legible translation of elementary environmental systems to local circumstances, this architectural work presents a beginning point and an anchor for this exhibition.

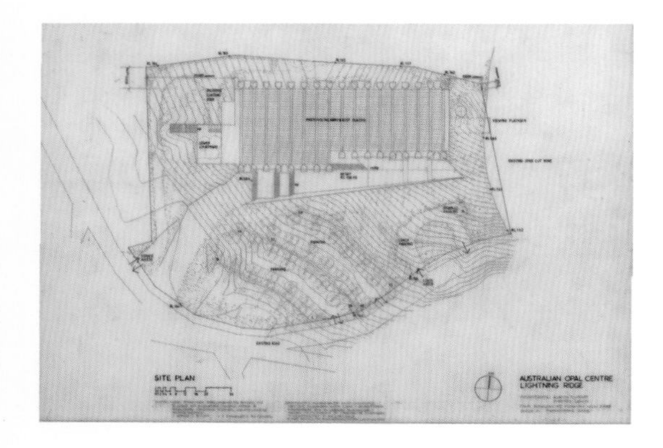

pp. 108–112: Original drawing by the architects. pp. 108–109: Long section. Opposite: Site plan. Opposite, above: Level 2 plan. Opposite, below: Level 1 plan. p. 112: Detail of the staircase and toplight. p. 113: Physical model of the Australian Opal Centre. Image by courtesy of the Architect.

第 108-112 页：建筑师的原始手稿。第 108-109 页：纵向剖面图。本页：总平面图手绘图纸资料。对页，上：二层平面图手绘图纸资料；对页，下：一层平面图手绘图纸资料。 第 112 页：楼梯及顶部采光细部图。 第 113 页：澳大利亚欧泊中心实体模型照片。

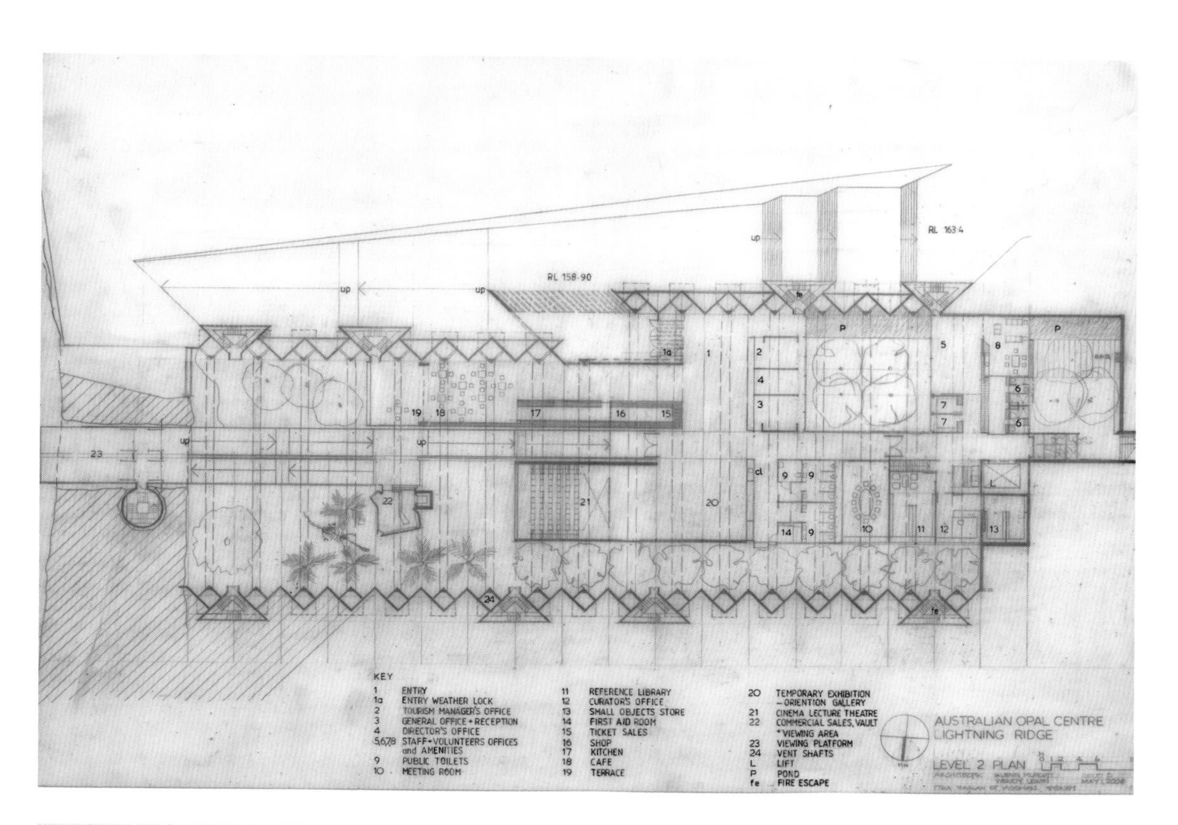
RL 163·4
RL 158·90
up
KEY
1 ENTRY
1a ENTRY WEATHER LOCK
2 TOURISM MANAGER'S OFFICE
3 GENERAL OFFICE + RECEPTION
4 DIRECTOR'S OFFICE
5,6,7,8 STAFF + VOLUNTEERS OFFICES and AMENITIES
9 PUBLIC TOILETS
10 MEETING ROOM
11 REFERENCE LIBRARY
12 CURATOR'S OFFICE
13 SMALL OBJECTS STORE
14 FIRST AID ROOM
15 TICKET SALES
16 SHOP
17 KITCHEN
18 CAFE
19 TERRACE
20 TEMPORARY EXHIBITION – ORIENTION GALLERY
21 CINEMA LECTURE THEATRE
22 COMMERCIAL SALES, VAULT + VIEWING AREA
23 VIEWING PLATFORM
24 VENT SHAFTS
L LIFT
P POND
fe FIRE ESCAPE
AUSTRALIAN OPAL CENTRE
LIGHTNING RIDGE
LEVEL 2 PLAN

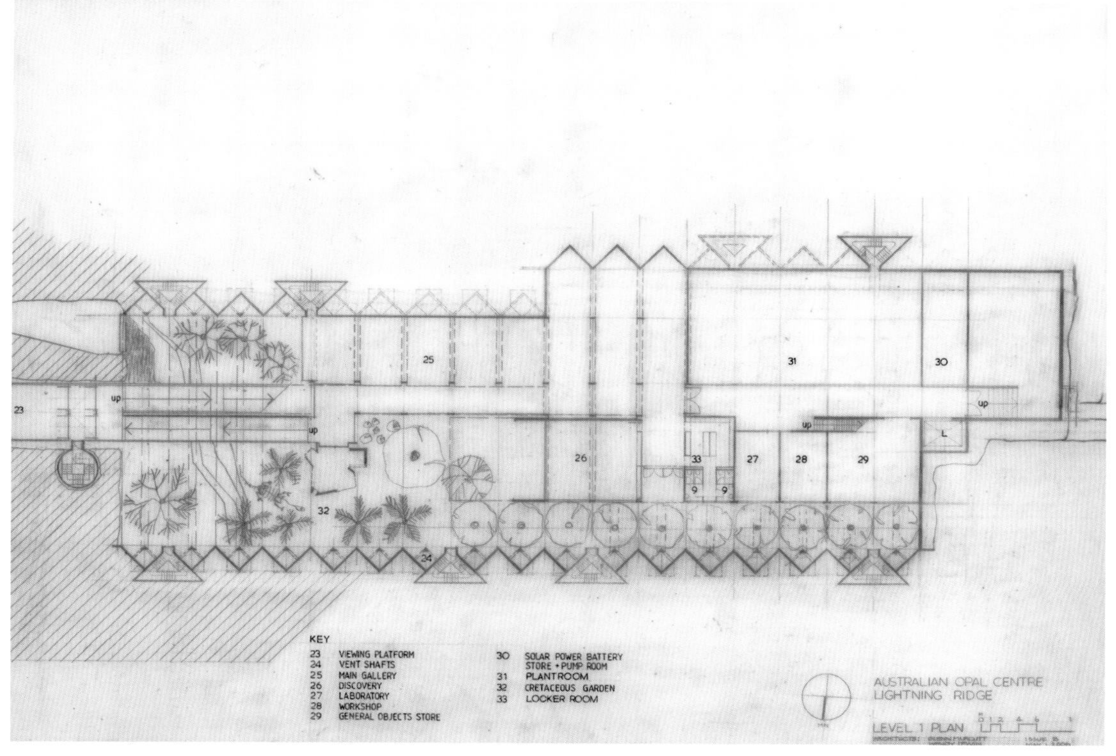
up
KEY
23 VIEWING PLATFORM
24 VENT SHAFTS
25 MAIN GALLERY
26 DISCOVERY
27 LABORATORY
28 WORKSHOP
29 GENERAL OBJECTS STORE
30 SOLAR POWER BATTERY STORE + PUMP ROOM
31 PLANTROOM
32 CRETACEOUS GARDEN
33 LOCKER ROOM
AUSTRALIAN OPAL CENTRE
LIGHTNING RIDGE
LEVEL 1 PLAN

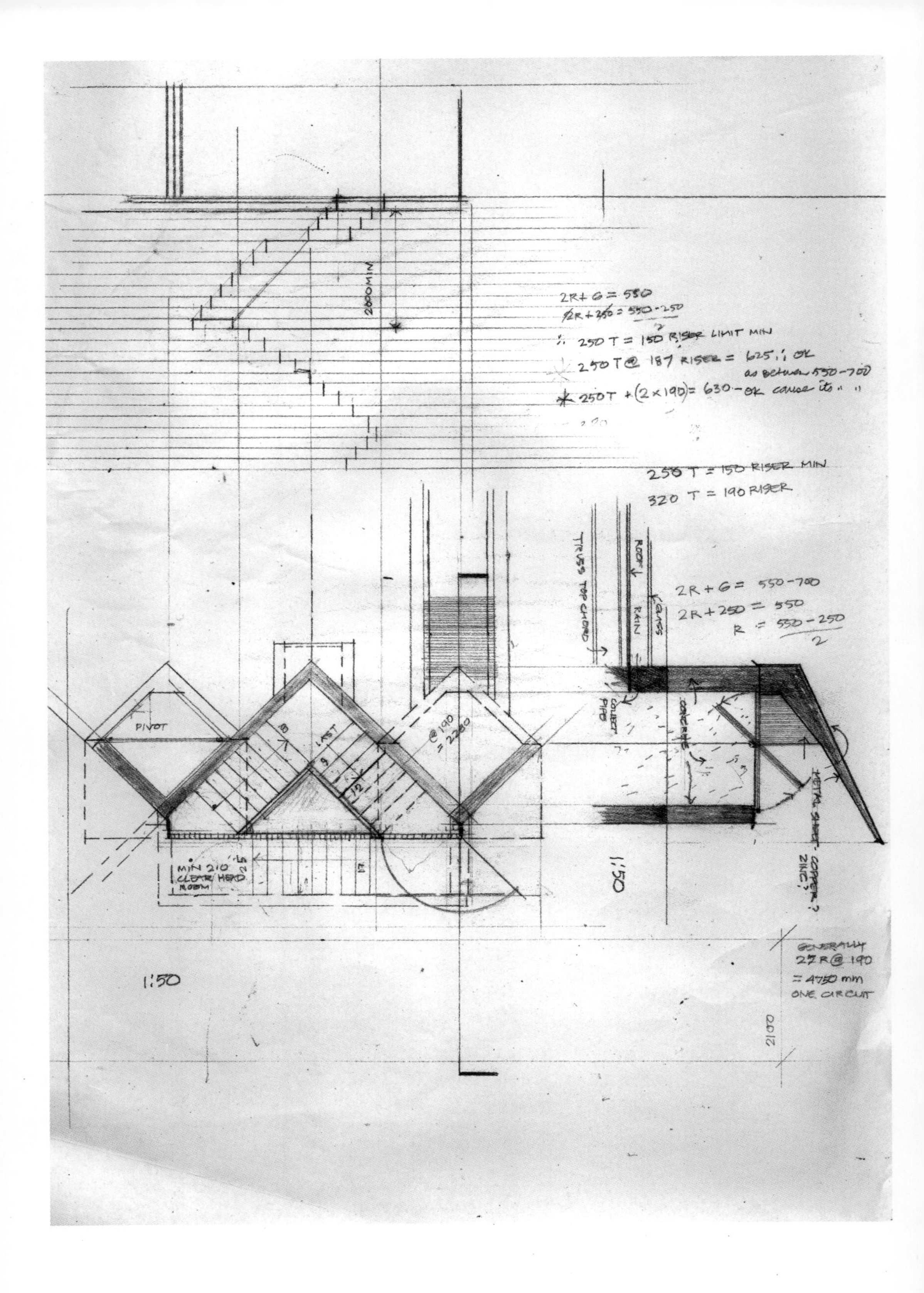
2000 MIN
2R+G = 550
2R+250 = 550-250
∴ 250 T = 150 RISER LIMIT MIN
250 T @ 187 RISER = 625 ∴ OK
as between 550-700
250 T + (2×190) = 630 - OK cause its " "
250 T = 150 RISER MIN
320 T = 190 RISER
TRUSS TOP CHORD
ROOF
RAIN
GLASS
2R + G = 550-700
2R + 250 = 550
R = 550-250 / 2
COLLECT PIPE
CONCRETE
PIVOT
LAST
@190 = 2280
METAL SHEET
COPPER? ZINC?
MIN 210 CLEAR HEAD ROOM
1:50
1:50
GENERALLY 27 R @ 190 = 4750 mm ONE CIRCUIT
2100

拟建的澳大利亚欧泊中心（以下简称 AOC）是一处研究兼教育设施，位于闪电岭——世界上黑欧泊（黑色蛋白石）出产量最高的标志性采矿镇。场地与外界隔绝，也没有任何城镇服务，与场地毗连的是历史悠久的 3 英里（约 4.8 千米）露天矿井，也是该地区开采量最大的矿井。在这个半干旱地区，夏季气温可升至 40° C 左右，冬季气温可降至 5°C 以下。

自 2003 年开发以来，在当地政府和闪电岭社区的积极参与下，AOC 将助力这个澳大利亚经济最落后的地区的文化和经济发展。它被构想为一个自主项目：将为自己供电，支持水的收集、存储和循环利用，拥有现场废物处理系统，并且融入被动式制热和制冷系统。它基本嵌入地下，利用地热和 22°C 地下恒温，减轻建筑的外部热负荷。马尔卡夫（一种捕风装置）利用了源自古代中东沙漠建筑的被动降温系统，从盛行风中吸取空气，并顺着烟囱向下导流，经过一系列湿板和水体，为室内提供凉爽且经过温湿度调节的空气。外部表达上，他们赋予建筑富有韵律和纪念意义的外观。将现场开挖出的材料混合到混凝土中，使得建筑物既存在于场地上，又成为场地的一部分。

在内部，一个平缓的斜坡将游客引导至悬浮在露天矿井上方的平台，参观者因此得以看到古老的地质层。下层可举办澳大利亚自然和文化宝藏地下展，囊括了全球最重要的稀有澳大利亚欧泊和世界各地的欧泊化石收藏品。展出内容还延伸至采矿机械，以及闪电岭的社会历史。内部设有一个花园，种植着源自冈瓦纳时期的古老植物，例如瓦勒迈杉、蕨类植物和苏铁植物，它们堪称植物活化石。

通过将基本环境系统直接且清晰地转译为当地情况，该建筑作品展示了本次展览的切入点和定位。

Essay:

System, Ambience, Translation

Maryam Gusheh and Philip Oldfield

论文：

系统，环境，转译

玛利亚姆·古谢，菲利普·欧菲尔德

In their recent account of the proposed Australian Opal Centre (AOC, See pp. 108-113) in Lightning Ridge, filmmakers Catherine Hunter and Bruce Inglis begin with a meditative aerial view of the surrounding landscape. Their camera flies over the dusty field dotted with soft sand-dunes and resilient native shrubs to arrive at the active underground mines that circle the townscape. Rudimentary steel shelters, interlocked with mechanical pulleys, attendant trucks and machinery, mark the point of descent below the ground's surface and signify the invisible human toil that takes place within these confined cavities. Immediately conveyed in this visual sequence is the harsh physicality of this environment and its lived history.

A perspective rendering depicts the design response to this enigmatic setting. The building is direct and elemental. A deep, expansive roof, overlaid with a field of inclined solar panels, harvests the sun's heat and collects rainwater and condensation. Cascading steps define the vertical passage down into the more temperate subterranean climate formed by the substantially embedded building. Wind towers, resolved as triangular prisms, give syncopated definition to the northern boundary wall and express catchment and flow of air down into the folds of the interior. The predominantly reinforced concrete material pallet integrates excavated rock as aggregate and imbues the material with a local texture and pigmentation. And deep within the building will be a garden supporting ancient vegetation. In its conception, expression and form, the architecture is explicitly engaged with the site, its associated patterns of use and the immediate natural phenomena. Resisting the broad banner of sustainability, architects Wendy Lewin and Glenn Murcutt describe the work instead as a partnership with the environment, a mode of architecture that is "in and of the site".

Invited to curate a display of contemporary Australian architecture with the AOC as an anchor point, Lewin describes her approach through the dialectical title, Universal Principles: Unique Projects[1]. Considered in light of priorities implicit in the AOC, and through the collated projects featured in this book, several intertwined attitudes to the intersection of architecture, environment and technology are resonant. We consider these themes with episodic reference to the featured projects. Our interest is not to categorise or to imply easy classification, for each and every work can readily traverse across and intersect our thematic structure. Our emphasis is instead on the breadth of architectural positions, on evocative alignments as well as distinct traditions.

System

A synthesised, almost synonymous relationship between fundamental architectural elements and environmental operations is emphasised. With an insistence on disciplinary literacy, craft and conventions, environmental limits or objectives are framed as the means by which architectural parts, components and entire building types are revitalised.

Designed by Ingenhoven Architects + Architectus, 1 Bligh Street (See pp. 156–167) uses technology as a fundamental driver to design. To maximise harbour views to the north, the core is pushed to the south. Then, a central atrium provides natural light and ventilation to deeper floor spaces. This acts in a similar manner to pitched roofscapes or wind towers of vernacular architecture, using the stack effect and buoyancy of air to drive ventilation, and provide, in this

在最近给闪电岭澳大利亚欧泊中心(AOC，见108-113页)拍摄的影像记录中，制片人凯瑟琳·亨特和布鲁斯·英格利斯用一场引人联想的俯拍拉开了序幕。镜头飞过尘土飞扬的原野，松软沙丘和柔韧灌木点缀其间，最终抵达小镇外围开采中的地下矿井。简陋的钢棚与机械滑轮、随行卡车、机器彼此交错，标志着地表下方的入口，也暗示着密闭矿洞中的人类劳作。严酷的环境和当地的居住史，便由这样的视觉顺序即刻传达给了观众。

一幅透视渲染图解密了这一场所的设计。建筑棱角分明，样式规矩。深邃宽阔的屋顶上铺着倾斜的太阳能板，转化太阳热量的同时收集雨水和凝结水。建筑大部分埋入地底，沿着台阶向下，气候随之变得温和。一连串三棱柱状的风塔切分了北面边界墙，空气可以沿着内壁循环流动。建材上主要使用了钢筋混凝土，而现场挖掘出土的岩石则作为骨料，给建筑添上一丝本土的纹理和色彩。在建筑的深处，还有一个培育古代植被的花园。从概念、表达到形式，整个建筑显然考虑了如何协调自然、场所及其使用形态。建筑师温蒂·卢因和格伦·马库特拒绝打出可持续性的旗号，而是更愿意将该作品形容为环境的同伴、一种“与场所同在”的建筑。

卢因受邀策划了一场以AOC为切入点的澳大利亚当代建筑展，而她的设计理念，也反映在辩证式标题之中——“普遍原理：特殊项目”[1]。通过了解AOC项目在设计时优先考虑的因素，以及本书关注的数个项目，可以看到建筑、环境、技术三个主题不同程度的互融。因此我们决定根据这些主题介绍项目。但我们的旨趣不在于进行简单的分门别类，毕竟每一个作品都或多或少地体现出不同主题的交融；我们强调的是宽泛的建筑观，是一以贯之的本质以及澳大利亚与众不同的传统。

系统

显而易见，建筑基本要素正在逐渐与环境性能合二为一，甚至接近同义。基于已有的专业知识、技能和传统，利用环境限制，或借助环境目标，可以重新理解建筑的部件和整体的类型。

英格霍芬建筑师事务所负责的布莱街1号(见156-167页)就是用技术驱动设计的例子。为了最大限度利用北面的海港景观，核心区被置于南面。中庭则给深邃的楼内空间提供了自然光和自然风，效果类似民居建筑里的斜屋顶或风塔，既能利用空气的烟囱效应和浮力效应促进通风，又为地面的公共空间创造了舒适的环境。安德鲁·伯恩斯建筑师事务所设计的克兰布鲁克学校(见120-129页)地处环境极其特殊的沃根河谷，恰好体现了环境、舒适度和使用者社交活动之间的关系。每天清晨，每个寄宿小屋都要派出四名学生取来柴火，燃起壁炉，为所有人供暖、烧水。如此一来，在建筑和系统的要求下，学生自然而然地为了集体利益参与合作、接触环境。事实上，从都市到家庭，都能找到这样的理念。以都市为例，让·努维尔建筑师事务所和PTW建筑师事务所共同设计的中央公园一号(见190-197页)中，30,000平方米的垂直绿化与建筑融为一体，使之显得生机勃勃，这在高层住宅中并不常见。不仅如此，绿植还减少了立面35%的热负荷，绿植的灌溉也完全依靠循环水。家庭层面上，群作工作室开发的节水浴室和移动式超级厨房(见54-59页)结合了原本应用于卫生设施、移动设备、灰水回收的工业技术，以及专门的工艺和技能，计划给未来住宅提供灵活的服务。同样地，特罗普建筑师事务所和奥斯卡建筑的合作项目“罗布特罗普分离舱”(Robe - trop_pods)(见60-67页)利用预制优势，打造出既自成一体又能随气候和环境

instance, comfort to the public spaces at ground level. In the very different context of the Wolgan Valley, Cranbrook School (See pp. 120–129) by Andrew Burns Architecture expresses the relationship between the environment, comfort and the social activities of its users. Four students from each lodge are expected to collect firewood early each morning to fuel slow combustion fireplaces for heating and hot water for the benefit of the whole group. The architecture and systems require the students to collaborate, and to engage with the landscape for their collective well-being. Such thinking is apparent from the urban scale to the domestic. In One Central Park (See pp. 190–197), by Ateliers Jean Nouvel and PTW, the radical integration of 30,000m² of vertical planting creates a vibrant architecture often absent from high-rise living. Yet it also reduces heat load on the façade by 35%, while relying entirely on recycled water production for irrigation. At the domestic scale, Crowd Production's water-conserving bathroom and mobile hyper kitchen (See pp. 54-59) combine industrial technologies for sanitation, mobility and grey-water recycling with craft and skill to create flexible, adaptable services for future habitation. Likewise, Troppo Architects and Oscar Building's trop_pods (See pp. 60-67) use the benefits of prefabrication to create an architecture of adaptable rooms and buildings, which is simultaneously systematic and engaging with the climate and environment.

Privileged across these projects is the radical reduction of architectural elements to the essential expression of their utility – at once operative and abstract. Two distinct yet interrelated traditions are pertinent here: the reinterpretation of indigenous, traditional and vernacular knowledge and ingenuity on one hand and the ongoing search for a revitalised and more elevated modernist vocabulary on the other.

Ambience

While material shelter remains fundamental to the architectural paradigm, Nature can creatively inform architecture. The collated projects in this book then seek to resist the wilful subjugation or control of the natural environment, and instead embrace its ambience. Natural phenomena – such as the sun, rain and the wind, fog, precipitation and vegetation - are deployed as active and at times unpredictable actors within the built environment. Both weather and weathering are celebrated. And with the acute attention to the evolution of architectural form in time, variations and inconsistencies, roughness in finish, natural over-growth and incursions are not only tolerated, but actively embraced. The architectural shelter is thus to simultaneously moderate and emphasise natural elements as agents in the shaping of the architecture and spatial experience.Mooloomba House (See pp. 76-83), by Andresen O' Gorman Architects, both emulates and harbours its context of Banksia forest, a raised linear home built around mature trees using local timber. The moderate coastal climate of South East Queensland is embraced through a design that encompasses significant outdoor, yet sheltered, rooms and spaces, providing an opportunity to live very immediately with the environment. Similarly, Richard Leplastrier's Lovett Bay House (See pp.84-91), on the fringe of Sydney's Ku-ring-gai Chase National Park, deploys the immediate topography as its outer boundary. The house is described as a room within a "greater room", with the surrounding cliffs the walls, and the floor the shifting tidal level of the river below. The design rejects a rigid boundary condition - and glass as a material entirely - instead using multilayered operable facades of plywood and fabric. These fold up and attach to the oversailing roof to allow for adjustments in enclosure, light, view and ventilation. Baracco and Wright Architects' Garden House (See pp. 44-53) goes further still, blurring the boundaries between the environment and the architecture, creating a building the architects describe as a "semi-permanent tent". Stripped back to the bare essentials, the house's permeable polycarbonate skin allows nature to seep under the walls, into the floorplan and overhead to present the building as a dynamic convergence between the architecture and landscape.

With the Shearers Quarters (See pp. 22-43), designed by John Wardle Architects, the exterior enclosure is clad in galvanised corrugated iron, a reflection of the likely materiality of the original shearing shed on the site. Yet the interior has a warmer material ambience, with Pinus macrocarpa in the living spaces, and recycled apple-box crates in the bedrooms, sourced from local orchards. Operable walls mediate inside and out to permeate this precise and beautifully crafted enclosure with atmospheric drama and fluctuation. Timber is also celebrated as a material in International

变化调整空间的建筑。

这些项目最突出的特点，是将建筑元素彻底简化为各种具体或抽象的功能。其中包括两个截然不同但又相互关联的传统：一方面是重新诠释本土自古以来的知识和智慧，另一方面是不断探索如何复兴和升华现代语言。

环境

建筑归根结底是一种庇护，而自然气候会给建筑带来创造性的影响。本书收录的项目在面对自然时，都选择用接纳的态度替代征服和控制。阳光、雨水、风、雾、降水、植被生长等自然现象皆被视为建成环境中积极且时而不可预测的因素，天气变化以及由此带来的自然风化也无需回避。设计者还意识到建筑形式会随时间演化，因而对于一些改变、差异、粗糙涂装、植被的过度生长和入侵等等，予以了包容。如果说自然元素是塑造建筑和空间体验的力量，那么作为遮蔽物的建筑就能够用来抑制或强化这种力量。安德烈森-奥格曼建筑师事务所设计的穆鲁姆巴住宅（见76-83页）是一座利用当地木材建造的线型住宅，与紧邻着的班克木林几乎融为一体。整个设计，包括重要的户外空间（尽管有遮蔽）和室内空间，都提供了与环境亲密共生的机会，住户可以充分感受昆士兰东南部沿海的温和气候。相似的项目还有理查德·莱普雷斯特里尔设计的洛维特湾住宅（见84-91页），位于悉尼库灵盖狩猎地国家公园的边缘，地形即对应其外部轮廓。它被形容为一个被"更大的房间"包裹着的房间，周围悬崖充当墙壁，下方因潮汐而变化的河面则是地板。室内和室外之间没有彻底的隔断，设计者拒绝用玻璃密封空间，而是用胶合板和织物组合出了可开合的多层次立面。这些立面可以折叠附着在悬挑屋顶上，以此调节室内的开放度、视野范围、采光和通风。巴拉科与赖特建筑师事务所的花园住宅（见44-53页）则更进一步：它模糊了环境和建筑之间的界限，被建筑师称为"半永久帐篷"。这座住宅简化到只剩下核心要素，透明的聚碳酸酯表皮使得自然因素可以从四周进入室内，进而使房屋成为建筑和景观的动态融合。

约翰·沃德尔建筑师事务所设计的剪羊毛工人的宿舍（见22-43页）外墙覆盖镀锌瓦楞铁板，延续了当地原有的剪羊毛棚的材料特性。相比之下，室内显得更加温馨——起居室使用了大果松，卧室则使用回收利用的苹果板条箱做内衬，可操控开合的墙面还让这个制作精良的空间呈现出戏剧般的流动感。由哲纳司设计的悉尼国际大厦（见168-179页），木材同样引人注目。事实上，对于多层商业建筑来说，木材作为主要结构材料不仅能降低隐含能耗（译注：指产品在制造、运输等生产链中造成的能源消耗），还会为内部的商业环境增添松木的自然色彩和质感，以及意想不到的木质香气。

转译

本书介绍的建筑类型和方案看似多样，实则反映了建筑与澳大利亚文化之间广泛的交集。每一个项目都会根据当地环境，调整常规的建筑策略，在项目推进中打磨出独一无二的设计。而产生影响的因素很多，从建筑方案的特性、生产的经济性，到根据城市密度和运作方式制定的监管框架，从教育和个人伦理，到相似群体与异质混合的协同效应。每个案例中，可见和不可见的变量都会引出新的解决方案。如果说这里强调的"地方性"有什么意义，那么便是转译环境带来的创造性发现。

在悉尼中央公园的案例中，都市设计者顶住了政治压力与舆论，未将城市内部的居住密度提升至曼哈顿、巴塞罗那的程度，而是确保大量资金用于改善周边环境、建造公共设施。即使是中央公园啤酒工业园（见180-189页），也在遗产保护条例的基础上满足了环境的要求。设计师哲纳司对工厂做了细致而大胆的改造，把重要的工业遗产转变为区域内供电、供暖、制冷的中心，当地一贯的适应性改造与修复的方式也由此受到冲击并得到改进。公园内的地标项目是中央公园一号，由让·努维尔建筑师事务所和PTW建筑师事务所设计。他们把乡村技术用于都市，全新的建筑表现随之诞生。同样被地区管理条例影响的，还有格伦·马库特的唐纳森之家（见92-107页）。为了符合易发火情的澳大利亚的严格标准，建筑师对他擅长的多层外皮设计进行调整，将通常用于防虫、遮光、防风的装置，改成了一个由玻璃、防火丝网和防火卷帘组成的三段式防护罩。立面上还开有最大直径为2毫米的小孔，开孔率达19%，呈现出新奇的视觉效果。除此之外，还有这样的一些案例：巴德·布兰尼根建筑师事务所负责的远在北昆士兰的莱斯·威尔逊澳洲肺鱼探索中心（见130-137页）出于公众利益，采用了一种合理的骨架式结构；西澳大利亚矿业小镇纽曼的东皮尔布拉艺术中心（见138-143页）由奥芙瑟·佐兹建筑师事务所设计

House (See pp. 168-179), Sydney, designed by Tzannes. Yet here, radically for a multi-storey commercial building, timber forms the primary structural material, benefiting from low embodied carbon, but also infusing the commercial interiors with natural colours, textures and the unexpected smell of pine.

Translation

A diversity of building types and programs in this book present a broad intersection of architecture and Australian culture. In each case, generic architectural strategies are adjusted to the particularities of the local environment and in the process uniquely transformed - from the specificities of the architectural program and the economics of production to the regulatory frameworks and codes that prescribe urban densities and operations, from educational pedagogies and personal ethics to collaborative and hybrid synergies. In each case, visible and invisible parameters precipitate new solutions. If there is a sense or a definition of the 'local' that is emphasised here, it is the inventive discoveries that emerge through contextual translations.

The urban design of Central Park in Sydney resolutely defied political pressure and controversy to translate inner city residential densities to those approximating Manhattan and Barcelona, while securing significant investment in environmental performance and public amenity. The Brewery Yard (See pp. 180-189), at the heart of this precinct, meets environmental ambitions in the context of heritage fabric and regulation. In this sensitive yet bold conversion of significant industrial heritage to a district power, heating and cooling plant, by Tzannes, the local conventions of adaptive reuse and restoration are challenged and elevated. At One Central Park - the iconic development within this precinct, by Ateliers Jean Nouvel and PTW Architects - the transfer of technologies from rural to the urban realm informs a novel architectural expression. Local codes are also at play at the H&P Donaldson House (See pp. 92-107) by Glenn Murcutt. The architect's familiar trope of a layered outer skin is here adapted to meet the stringent Australian Standards for high-risk fire-prone environments. A device typically used for the protection against insects and mediation of light and air is thus translated as a three-partite fire shield consisting of glazing, mesh fire screen and fire shutter. The maximum allowable perforations of 2mm diameter holes at 19% of the surface area result in a new figural emphasis. At the Les Wilson Barramundi Discovery Centre (See pp. 130-137) in far North Queensland, designed by Bud Brannigan Architects, a rationally resolved skeletal structure is invested with civic purpose; while at the East Pilbara Arts Centre (See pp. 138-143) in the Western Australian mining town of Newman, by Officer Woods Architects, an overscaled industrial envelope affords both civic permanence and adaptability. CHROFI's design for the Ian Potter Foundation National Conservatory (See pp. 144-151) at the Australian National Botanic Gardens in Canberra reinterprets the typology of the glasshouse. The design instils what is a traditionally lightweight structure with thermal mass for performance, but also adapts light, transparency and program to reinvent the spectacle of the glasshouse for the visitors. And design-led research, as part of the Cooperative Research for Water Sensitive Cities (See pp. 198-209), draws on local environmental history to propose new productive partnerships between urban and natural environments. At once restorative and transforming, these projects reveal intrinsic characteristics, address immediate challenges and imagine more humane and robust futures for our cities.

A Situated Ethic

In Australia, as with other regions outside the dominant centres of modernist thought, nature, landscape and the weather are recurring motifs in the search for the local, for what should inform the local architectural character and approach. Critics attuned to nineteenth-century colonial stereotypes have countered the sentiment. The synonymous reading of nature and culture, it has been pointed out, diminishes the local culture and intellect to a point where only climatic conditions remain. The collection of architectural projects represented in this issue surpasses such reductions to propose environmental relationships informed by and embedded in local history, society and culture. In their diversity, they convey a variety of attitudes to an empathetic engagement with the natural setting, encompassing transience, adaptive reuse, technological performance, prefabrication and operational autonomy. It is proposed that the common ground across this collective is an innovative synthesis of architectural principles and a situated ethic.

以超大规模的工业外观为公众提供了永久、灵活的空间；克洛菲设计的伊恩·波特国家温室（见144-151页）位于堪培拉澳大利亚国家植物园内，重新诠释了温室的形态，融入带蓄热块的传统轻质结构，调节光线和透明度，为参观者展现温室本身的梦幻；澳大利亚的水城市主义（见198-209页）则以设计为主导、以本地环境史为参照，提出在城市和自然环境之间建立卓有成效的新型伙伴关系。这些项目既是修复也是变革，揭示了城市的本质，指出了眼前的挑战，为我们的城市设想了更有人情味也更稳健的未来。

情境伦理

与那些现代主义思想影响范围之外的边缘地区一样，若在澳大利亚寻找地方性，寻找影响当地建筑外观和建造方法的要素，自然、景观、天气这样的母题必将反复出现。然而，习惯了19世纪刻板的殖民样式的批评家对此无法表示赞同。他们指出，将自然与文化画上等号的解读似乎是暗指当地只剩下气候条件堪为特色，这是对当地文化和智慧的贬低。本书集合的项目设计者没有止步于这种浅薄的理解，而是认为建筑与环境的关系根植于本土历史、社会、文化，且受其影响。这些项目的多元性，反映出设计者与自然共鸣的不同态度和方式：有的是短暂介入，有的是适应性改造，有的予以技术赋能，有的利用预制组件，还有的追求自主运营。但即使项目千差万别，建筑原理和情景伦理的融合仍是共同的立足之本。

Note:

1. Featured at the exhibition "Universal Principles: Unique Projects, Australian Architecture Re-Setting the Agenda." The exhibition was commissioned by the Australian Government's Department of Foreign Affairs and Trade and exhibited at Sky Gallery 3, Tokyo City View, 52 F, Mori Tower, 7 July–26 August 2018. Curator: Wendy Lewin. Curatorial Team: Maryam Gusheh (Australia), Tom Heneghan, Souhei Imamura and Hitomi Toku (Japan). Organising Committee: Wendy Lewin, Tom Heneghan, Souhei Imamura and Hitomi Toku.

注释：

1. 展览"普遍原理：特殊项目——澳大利亚建筑的重启"由澳大利亚政府的外交贸易部委托举办，地点在森大厦52层东京六本木新城展望台的天空画廊3，展期为2018年7月7日至8月26日。策展人：温蒂·卢因。策展团队：玛利亚姆·古谢（澳大利亚），汤姆·赫尼根，今村创平，德仁美（日本）。组委会：温蒂·卢因，汤姆·赫尼根，今村创平，德仁美。

Maryam Gusheh is Associate Professor in Architecture at Monash University.
Philip Oldfield is Director of the Architecture Program at UNSW, Sydney.

玛利亚姆·古谢 莫纳什大学建筑系助理教授
菲利普·欧菲尔德 悉尼新南威尔士大学建筑系主任

Andrew Burns Architecture

Cranbrook School Wolgan Valley Campus

Greater Blue Mountains National Park, New South Wales 2015–2017

安德鲁·伯恩斯建筑师事务所

克兰布鲁克学校沃根谷校区

新南威尔士州，蓝山国家公园 2015-2017

Located 180 km northwest of Sydney in the Greater Blue Mountains area, this extraordinary site in the Wolgan Valley is defined by dramatic sandstone escarpments, fertile paddocks and expansive clear skies. The campus has been established to extend the Cranbrook School's educational philosophy and interest in rituals of stewardship; encouraging students to explore and strengthen their skills and sense of self in a social and physical environment, well-removed from their normal frames of reference, and develop sensitivity, understanding, respect for and connection with the natural world. When completed, the campus will be experientially rich and accommodate a lodge and kitchen, a chapel and hall, internal and external teaching spaces, and five separate buildings, each with 20 sleeping spaces and an observatory. The siting of the pavilions supports group learning from the classrooms to the kitchen, collaborative skills and social responsibility.

The site is off the grid, requiring the campus to function as an autonomous entity. Structures are arranged along a crescentshaped path. The majority are positioned on the outside of the arc, oriented toward the escarpment. The communal spaces are situated on the inside of the arc. Utility services are embedded, tracing the arc and establishing a service spine to distribute power, water and drainage services. Operation of passive environmental systems is the daily responsibility of the students; for example, adjusting the shading and ventilation to manage thermal performance, which in turn supports and improves their comfort. Their collective efforts at gathering firewood from the site for the slow combustion stoves provides warmth in winter and hot water all year round. The building systems for water and solar power are also monitored by students, who must assess water storage levels and power usage and evaluate and prepare appropriate responses.

The built forms comprise simple external volumes with skillion roofs, and more complex, sculptured interiors simultaneously addressing the particularity of program, and referencing the non-specific form of the Australian vernacular. A deliberately "spare" approach to siting, tectonics and enclosure supports a heightened engagement with the Wolgan Valley landscape and its embedded natural processes by young individuals, and together, as a community.

Site plan (scale: 1/20,000)／总平面图(比例: 1/20,000)

p. 121: The building form is made up of simple volume with skillion roof. Opposite: Series of chimneys reflect remnant chimneys from the neighbouring historic town of Newnes. This page: General view of the site after its first stage of completion. All images on pp. 120–129 by courtesy of the Architect.

第 121 页：建筑形式由简单的形体和单坡屋顶构成。对页：这里的一系列烟囱是对邻近的历史名镇纽恩斯残存烟囱的反映。本页：现场第一阶段建设完成后的全景。

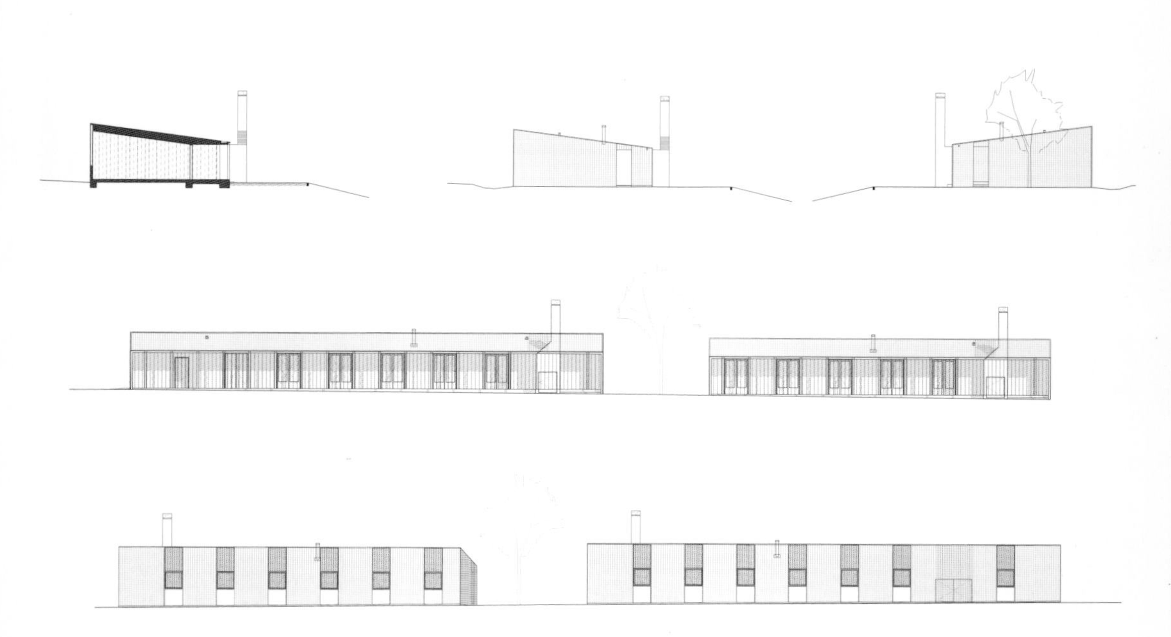

Typical section and elevations (scale: 1/500)／剖面图和立面图(比例: 1/500)

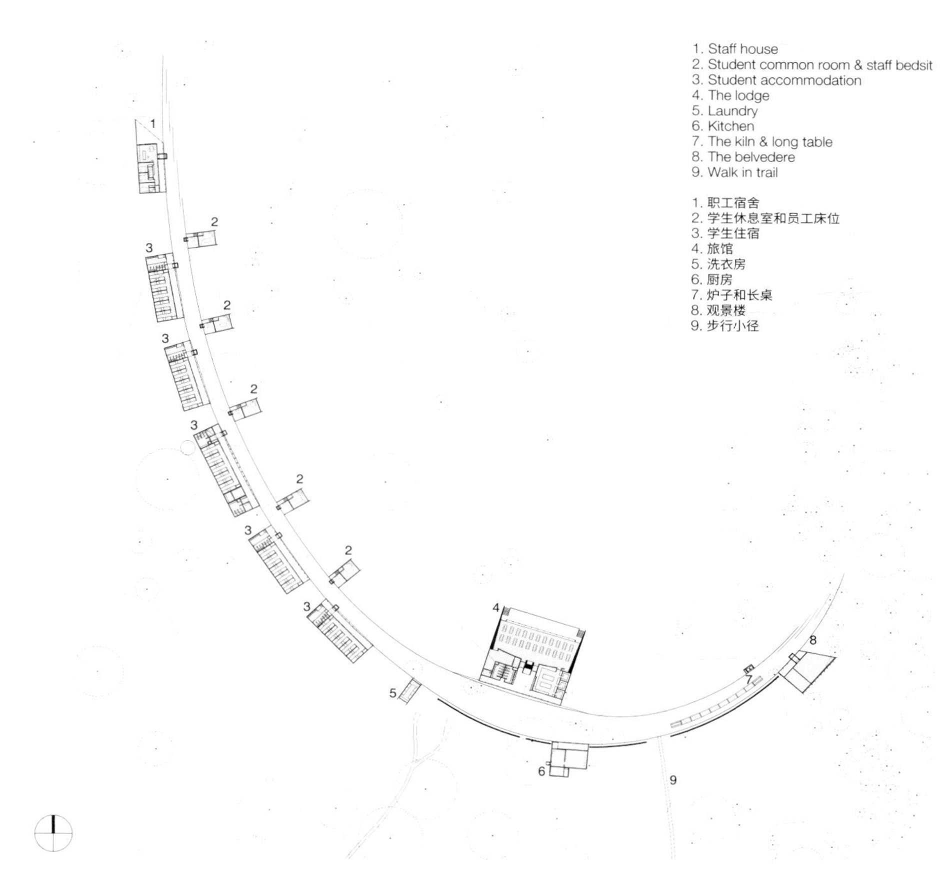

Plan (Scale: 1/2,500)／平面图(比例: 1/2,500)

沃根谷在距离悉尼 18 千米的西北方向，地处大蓝山山脉地区，以陡峭的砂岩绝壁、肥沃的牧场和一望无际的晴空而广为人知。校区建设是为了推广克兰布鲁克学校的教育理念及其对仪式化管理的兴趣；鼓励学生在社交和自然环境中探索和增强自己的技能和自我意识，远离常规的参照准则，同时提高对自然敏锐的感知力和理解力，学会尊重自然，并与自然建立联系。校园建成后将拥有丰富的体验，可容纳一间传达室和厨房、一个小教堂和礼堂，室内和室外教学空间，以及 5 栋单独的建筑，每栋均设有 20 处睡眠空间和 1 个瞭望台。这组建筑的选址既鼓励学生从教室到厨房都可以进行小组学习，又帮助学生培养协作能力和社会责任感。

该建筑基地未接入公用输电网、煤气输送网、自来水网等，因此需要校区成为一个能够自主运作的整体。建筑群沿着新月形的路径排布。大部分位于圆弧的外侧，朝向悬崖。公共空间位于圆弧内侧。公用设施沿着弧线被嵌入，并建立服务中心来配电、供水和排水。操作被动式环境系统是学生的日常工作；例如，调节遮阳和通风以控制热性能，进而维持并提升空间舒适性。他们共同努力，从场地收集慢火炉所需的柴火，保障全年的热水供应，并在冬季时为校区供暖。学生还监控建筑的水系统和太阳能系统，他们需要评估蓄水量和用电量，并根据评估做出合理应对。

建筑外部形式融合了简单的形体和单坡屋顶；而更为复杂、精雕细琢的室内空间在满足特定功能性的同时，也参照了澳大利亚乡土建筑的一般形式。选址、构造和围场上的有意“留白”，加强了年轻人及其组成的社区与沃根谷的互动，鼓励他们深入自然。

Credits and Data

Project title: Cranbrook School Wolgan Valley Campus
Client: Cranbrook School
Location: Greater Blue Mountains National Park, New South Wales, Australia
Design: 2015
Completion: 2017 (Stage 1), Ongoing (Stage 2)
Architect: Andrew Burns Architecture
Design Team: Andrew Burns, Casey Bryant, Noel Roche, Carter Hu, Kate Fife, Jonathon Donnelly, Paul Coppere
Consultants: Turf Design (Landscape Design), TTW (Civil & Structural Engineer), PDS Group (Project Manager), JHA (Services Engineer), Hines Constructions (Construction)
Project area: 600 m² (Stage 1), 1,900 m² (Stage 2)

p. 124: Aerial view of the site and landscape.pp. 126–127: Interior view of the teaching space. Opposite and this page: Both images showing views from the interior. The school is currently at its second construction stage.

第 124 页：建筑基地及景观鸟瞰图。第 126-127 页：教学空间内部。对页及本页：二者皆展示了从室内看到的景观。该校区目前处于第二期建设阶段。

Bud Brannigan Architects

Les Wilson Barramundi Discovery Centre

Karumba, Queensland 2017

巴德·布兰尼根建筑师事务所

莱斯·威尔逊澳洲肺鱼探索中心

昆士兰州，卡伦巴 2017

The Les Wilson Barramundi Discovery Centre, for the Carpentaria Shire Council, is in Karumba, a regional township with a population of 600. On the edge of the Gulf of Carpentaria in Far North Queensland, the location is remote — 800 km from Cairns on the state's east coast. The environmental conditions are extreme, with prolonged periods of high temperatures, humidity, heavy rainfall and cyclones. The salt-laden air accelerates material corrosion. Adjacent to the Norman River, the site has a limited built context and is highly legible within a broad savanna landscape.

A redevelopment of a visitor centre on a barramundi farm and hatchery, the architecture is deliberately direct and expressive of its expanded program. It builds on its past as a visitor centre, telling the story of the Southern Gulf barramundi, and supports a research facility.

A 130-m curvilinear plan embraces a 2,500-m² pond, containing several thousand young barramundi from the hatchery. Conceived as an elevated platform, the facility is raised one meter above ground to clear established flood levels, provide for drawing cool air into the building from below and, with additional measures, establish a point of entry secure from crocodiles. The selected materials and assembly system are congruent with and invite an appreciation of the harsh local environment. A series of prefabricated skeletal steel portals, on a regular radial grid, simply support secondary framing and thin external metal cladding. This direct approach gains nuance through the dimensional adjustments of the structural portals, together with the incline of the eastern columns. The result is a dynamic sequence in section, plan and volume, culminating in a 14-m high roof lantern over the entry verandah. The figural transformation of a simple extruded shed to a local marker eloquently resolves this complex program within its extreme environmental context, confirming the farm's local and regional significance.

Credits and Data

Project title: Les Wilson Barramundi Discovery Centre
Client: Carpentaria Shire Council
Location: Karumba, Queensland, Australia
Design: 2015 – 2016
Completion: December 2017
Architect: Bud Brannigan Architects
Design Team: Bud Brannigan, Duncan Maxwell, Melina Hobday
Consultants: Ross Argent (Structural Engineer), Webb Australia (Electrical and Mechanical Engineers), Gilboy and Associates (Hydraulics Engineers), Andrew Prowse (Landscape Architect), Brandi (Interpretive Designers), Davis Langdon (Quantity Surveyor), Wren Construction (Builder)
Project area: 1,600 m²
Project estimate: $5 million AUD, excluding external works

pp. 130–131: General view of the building suspended a metre from the ground facing the 2,500 m² growing pond. Opposite: Distant view. Photo taken during the construction of the building. This page: View from the continuous verandah along the pond, shaded by a perforated metal screen. All images on pp. 130–137 by David Sandison.

第130-131页：全景图。建筑高出地面1米，面向2,500平方米鱼塘。对页：远景照片，拍摄于建筑施工期间。本页：站在鱼塘边有金属网罩遮荫的连续游廊上所欣赏到的景色。

由卡彭塔里亚郡议会兴建的莱斯·威尔逊澳洲肺鱼探索中心位于卡伦巴镇，该镇是一个只有 600 人的区域城镇。该地位于远北昆士兰的卡彭塔里亚海湾边缘，地理位置偏远，与昆士兰州东海岸的凯恩斯相距 800 千米。这里环境条件极度恶劣，长期处于高温、潮湿、强降水和气旋的状态下。含盐空气加速了物质的腐蚀。基地毗邻诺曼河，周围人造环境极少，在广阔的热带稀树草原景观中，基地清晰可辨。

在澳洲肺鱼养殖和孵化场重建游客中心，建筑显得从容而直接，并极具表现力地展现了它的扩展功能。它建立在过去游客中心的基础上，讲述着南部海湾肺鱼的故事，并为研究机构提供支持。

长达 130 米的曲线形设计包含了一个 2,500 平方米的鱼塘，里面有数千尾来自孵化场的小肺鱼。该设施被构想为一个高架平台，高出地面 1 米，在既定洪水位高度内不构成任何阻碍；高架平台使冷空气可以从下方被吸入建筑内部。同时，建筑采用额外措施，设立了安全入口，防止鳄鱼进入。该项目所选材料和装配系统与恶劣的当地环境相称，并加深了人们对环境的理解。一系列预制的门式钢骨架排布在规则的径向网格上，支撑着次结构和外部薄薄的金属表皮。使用者可以通过调节门式钢骨架的尺寸和东面立柱的倾斜度来直接达成形态上的细微差别。通过这种方法，实现了一座在剖面、平面和体量上都不断发生有序变化的建筑。入口走廊上方一个 14 米高的日光顶棚使设计达到高潮。建筑师将一个简单的棚屋形象化为一个本地标志建筑，巧妙地实现了这个在极端环境下的复杂项目，从而确立了该养殖场在地方和区域的重要性。

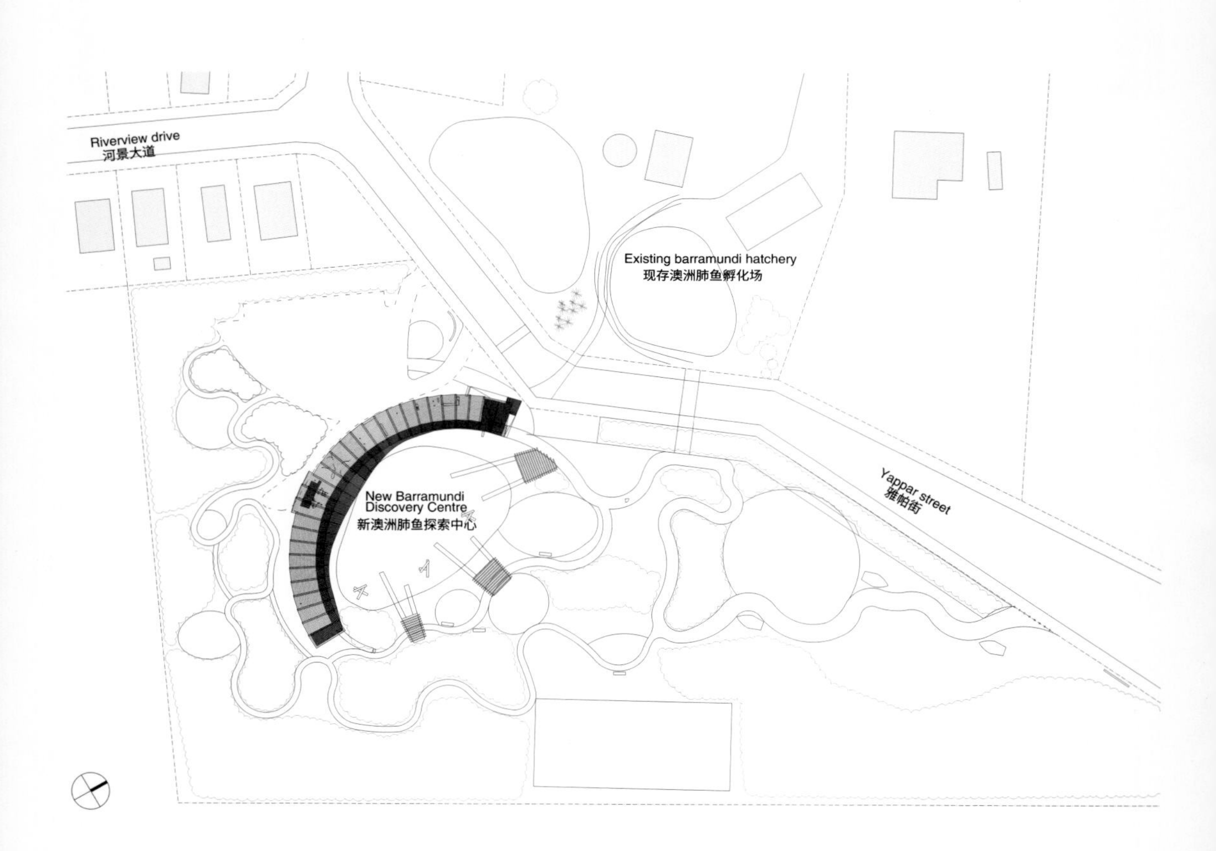

Site plan (scale: 1/2,500) ／总平面图（比例：1/2,500）

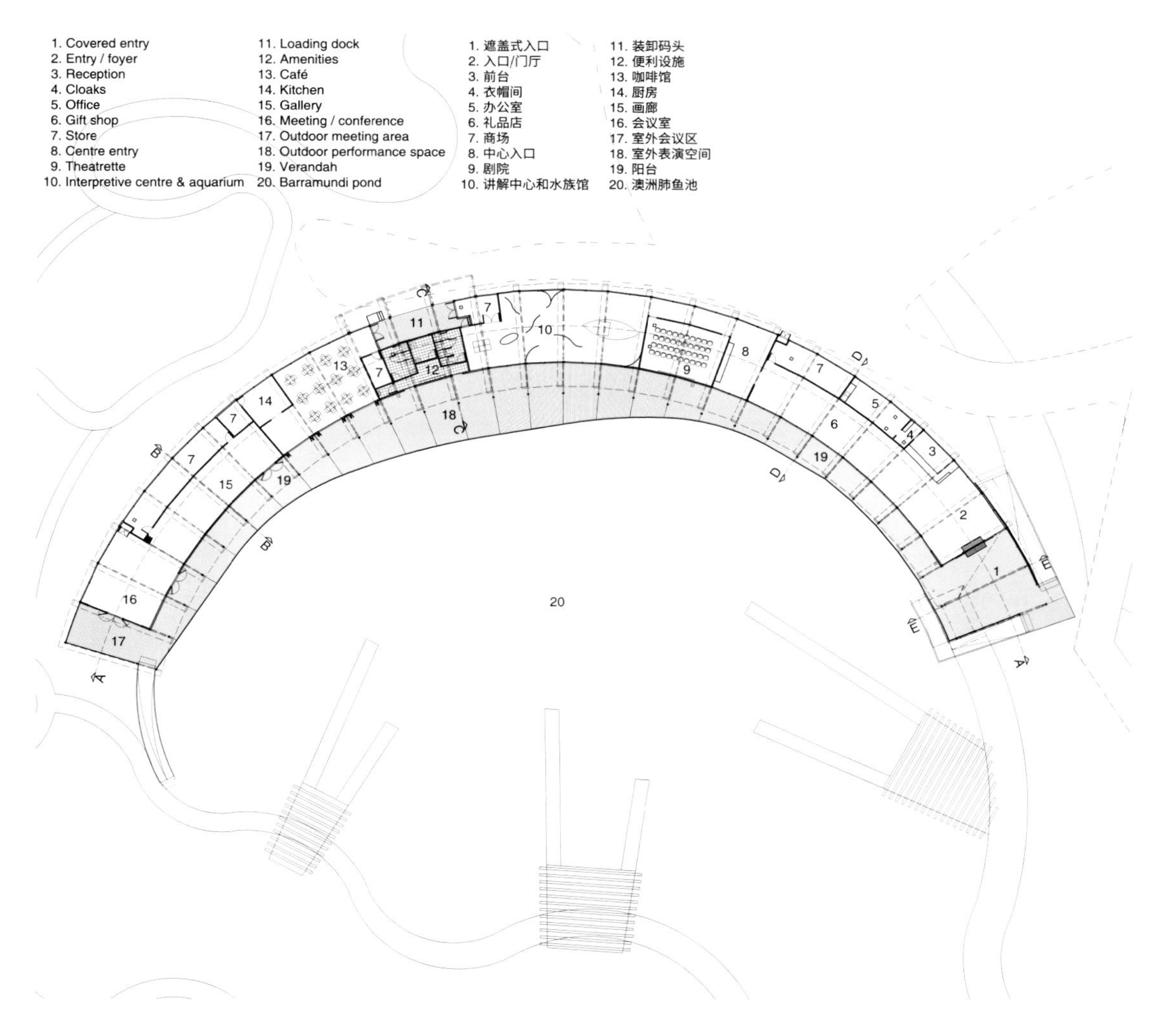

Plan (scale: 1/800) ／平面图(比例:1/800)

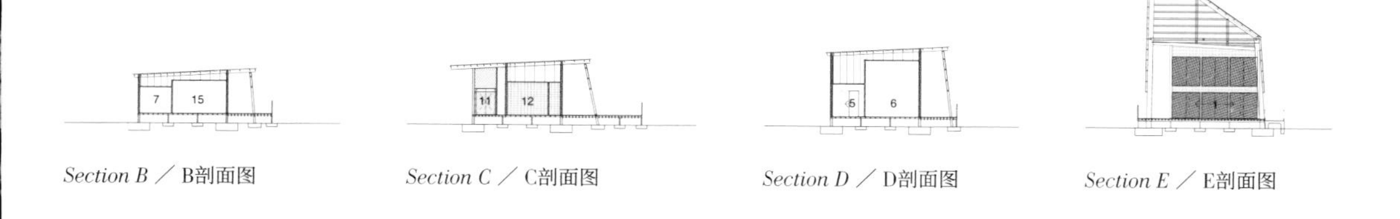

Section B ／ B剖面图　*Section C* ／ C剖面图　*Section D* ／ D剖面图　*Section E* ／ E剖面图

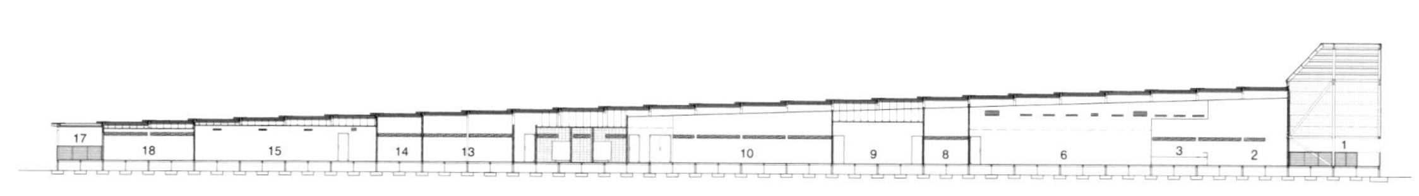

Section A (scale: 1/800) ／A剖面图(比例:1/800)

Opposite: Interior view looking towards the reception of the Discovery Centre. This page: Interior view along the verandah and pond.

对页：从室内望向探索中心接待处。本页：沿着游廊和鱼塘的室内空间。

Officer Woods Architects

East Pilbara Arts Centre

Newman, Western Australia 2016

奥菲瑟·伍兹建筑师事务所

东皮尔巴拉艺术中心

西澳大利亚州，纽曼 2016

FIRE
HOSE
REEL
EVENT
AHEAD

Credits and Data

Project title: East Pilbara Arts Centre
Client: Shire of East Pilbara
Location: Newman, Western Australia, Australia
Design: 2014
Completion: 2016
Architect: Officer Woods Architects
Design Team: Jennie Officer, Trent Woods, Melita Tomic, Oenone Rooksby, Monja Johnstone, Jack Choi
Consultants: Ralph Beattie Bosworth (Quantity Surveyor), Wood and Grieve Engineers (Civil, Structural, Mechanical), Best Electrical (Electrical), Schwanke Consulting (Fire and Compliance)
Project area: 1,945 m^2
Project estimate: $7.6 million AUD

The East Pilbara Arts Centre (EPAC) is a partnership between Martumili Artists, Shire of East Pilbara, BHP Billiton Iron Ore, the Pilbara Development Commission and Lotteries West. The building was innovatively procured through a design competition led by the University of Western Australia.

EPAC is a contemporary, flexible gallery and working space that provides Martumili Artists, an organization representing over 250 self-employed artists spread across seven Aboriginal communities in the eastern Pilbara, with a purpose-built creative facility for cultural expression, education and public exhibitions of work.

The design exceeds the scale of the brief by three times: allowing for future flexibility and anticipating further uses than those initially briefed. It meets the ambitions of a transformational building for Newman, a powerful demonstration of commitment to social, cultural and creative excellence and inclusion: both an exemplary gallery and a large flexible civic space that is the location of diverse appropriations and community events.

Designed as a large-span steel frame shed, thermally decoupled from programmed spaces within it, EPAC is both constructionally efficient and a durable casing around the gentle activities and accommodation related to the making, celebration and support of art and artists. The design is highly responsive to its site, with trees within the building, parking under the expansive roof and red dirt continuing under its skirt. Its scale and presence, announced by a 45-m long artwork, demonstrates its iconic civic nature, and it aims to generously lay itself open to its community and context. At any time of the year, occupants have ample space for temporary or permanent activity in and around the building, able to be contained, covered or completely open. A water tank located deep in shade services the building and provides a thermal sink. The planning exploits this passive system, encouraging artists to lean against the tanks that remain cool in the heat of the day, and simply find seclusion or separation.

Deeply encased, the gallery and support spaces use lowmaintenance and robust materials for unembellished but comfortable occupation. Utilizing the available double height of the main structure, the gallery is a voluminous, flexible space with a large operable wall. Visitors to the gallery may informally see artists at work as they move through the building, but the design provides artists seclusion if they wish. The design lets interaction happen spontaneously, rather than orchestrating a synthetic experience of viewing or being viewed.

pp. 138–139: The ventilated fly roof reduces heat load providing an effective response to the varied climatic conditions in its location. Opposite: Translucent panel lift doors provide multiple entry points and adjustable ventilation. All images on pp. 139-143 by courtesy of the Architect.

第138-139页：通风的悬浮屋顶可降低热负荷，有效应对其所处位置的各种气候条件。对页：使用半透明面板的升降门，可为建筑提供多个入口并调节通风。

东皮尔巴拉艺术中心（以下简称 EPAC）是一个合作项目，合作方包括马图米利艺术家、东皮尔巴拉郡、必和必拓铁矿、皮尔巴拉发展委员会和西部彩票。该建筑是通过西澳大学主导的设计竞赛，以创新的方式促成的。

EPAC 是一个灵活的当代画廊和工作空间，为马图米利艺术家（该组织代表了分布在皮尔巴拉东部 7 个原住民社区的 250 多名个体艺术家）打造了专属的创意设施，用于文化表达、教育和作品的公开展览。

该设计超出了任务书所述规模的 3 倍：考虑到未来的灵活性，预设了比原设计任务要求更加丰富的用途。它满足了纽曼改造建筑的雄心壮志，有力地证明了其对社会、文化、创意卓越性和包容性的承诺。它既是一个典范画廊，也是一个适用于多样用途和社区活动的大型灵活市民空间。

EPAC 的设计是一个大跨度钢架棚结构，钢架构与其功能空间实现了热隔离。高效、耐久的外壳包裹着活动空间和住宿空间，为创作、庆典及与艺术和艺术家相关的活动提供支持。该建筑的设计对场地作出了十分积极的响应。建筑内部有树木，宽阔的屋顶下设有停车位，踢脚板下方是连绵不断的红色泥土。这座建筑的规模与气质通过一件 45 米长的艺术作品得以彰显。这样的规模与气质展示着这座建筑服务于市民的本质，并致力于对其社区及环境开放。在一年中的任何时候，使用者都能在建筑内部或周围找到充足的空间进行临时或永久性的活动，空间或四周封闭，或天棚遮盖，抑或完全开放。常处阴影中的水罐既是建筑的供水设施，又是建筑的吸热装置。设计利用了这一被动式系统，鼓励艺术家们利用这些能在酷暑中保持凉爽的水罐，进行空间的隔离和划分。

建筑深处的画廊及其配套空间选用维护成本低且坚固的材料，打造了无装饰但舒适的使用体验。设计利用主体结构的双层挑高，使画廊成为一个空旷、灵活且拥有大型活动隔墙的空间。到访画廊的参观者们穿行于建筑中，还有机会偶遇正在创作的艺术家，但当艺术家不想被打扰时，建筑也设有能满足他们需求的清净空间。这一设计使交互自然而然地发生，而非刻意去安排一种“看”与“被看”的体验。

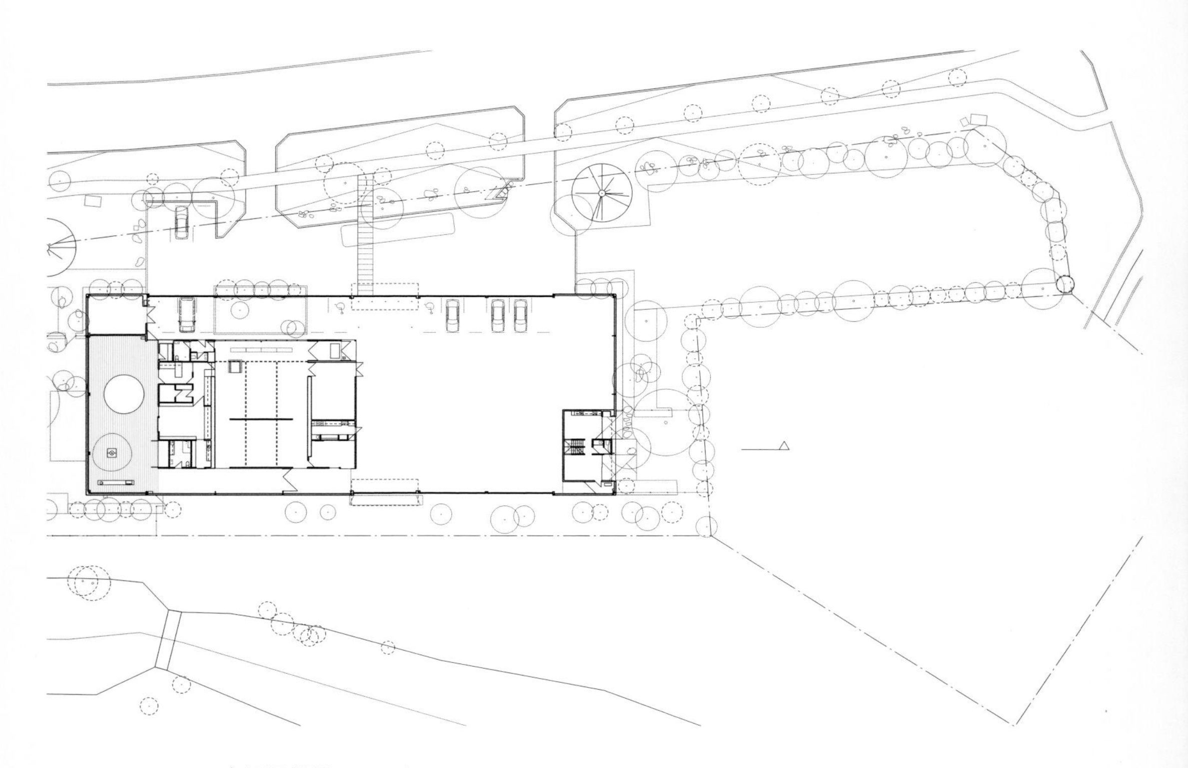

Plan (scale: 1/1,000) ／平面图（比例: 1/1,000）

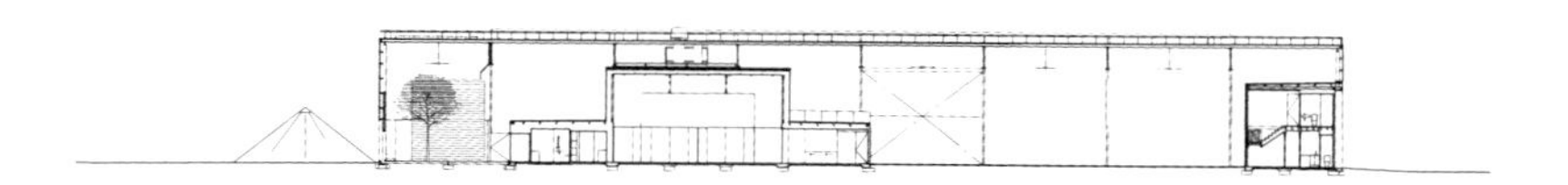

Section (scale: 1/1,000) ／剖面图(比例: 1/1,000)

This page: The storage water tank, an icon to the outback life, is used both as an architectural feature and as a passive cooling element in the artists' working area.

本页：储水罐是内陆地区生活的象征，既成为建筑特征，又用作艺术家工作区的被动式降温装置。

CHROFI

The Ian Potter National Conservatory
Canberra, Australian Capital Territory 2020

克洛菲
伊恩・波特国家温室
澳大利亚首都直辖区，堪培拉 2020

Credits and Data
Project title: The Ian Potter National Conservatory
Client: Australian National Botanic Gardens
Location: Canberra, Australian Capital Territory, Australia
Design: 2018
Completion: Expected completion 2020
Architect: CHROFI
Consultants: McGregor Coxall (Landscape and Plant Exhibit Design), Atelier 10 (Environmental Design), SDA Structures (Structural Engineer), Inhabit (Façade Engineer), Steensen Varming (Services Engineer)
Project area: 500 m²
Project estimate: $7.5 million AUD

The Ian Potter Foundation National Conservatory is located within the Australian National Botanic Gardens in Canberra, Australia's capital. The garden's national collection of native plants is significant, and has been curated for education, botanical research and cultivation. Lightly poised within this garden setting, this new center will support the display, conservation and propagation of rare and threatened tropical plants, and provide a key attraction for the gardens.

The design is informed by a central consideration: how can conservatories remain unique and memorable experiences in the age of climate change? This question is eloquently addressed in the context of Canberra's unaccommodating winter climate. Typically, the conservatory typology proposes light-framed, fully-glazed enclosures offering poor insulation during the winter cold, and exposing the interior to excessive heat gain in summer. This project proposes an approach that aligns the conservatory with the principles of botanical conservation, with the particularities in its immediate environment, and climate change.

A cubic volume appears to hover over the landscape and frames the sky. A central void provides the necessary light levels for the tropical plant species to thrive, while limiting the extent of external transparent surfaces. Surrounding this void is a layered skin, which acts as a high-performing insulating structure. Its outermost layer uses solar energy to pre-warm the air and store excess heat within the secondary wall thermal mass; it thus allows the building to maintain its high internal temperature during winter and mitigates high temperatures in summer. This outermost layer changes from transparent to translucent according to its orientation, creating a unique visual expression. The interplay of mass and light in this built work develops an atmospheric interior, and coalesces world-leading environmental performance with an elevated aesthetic experience.

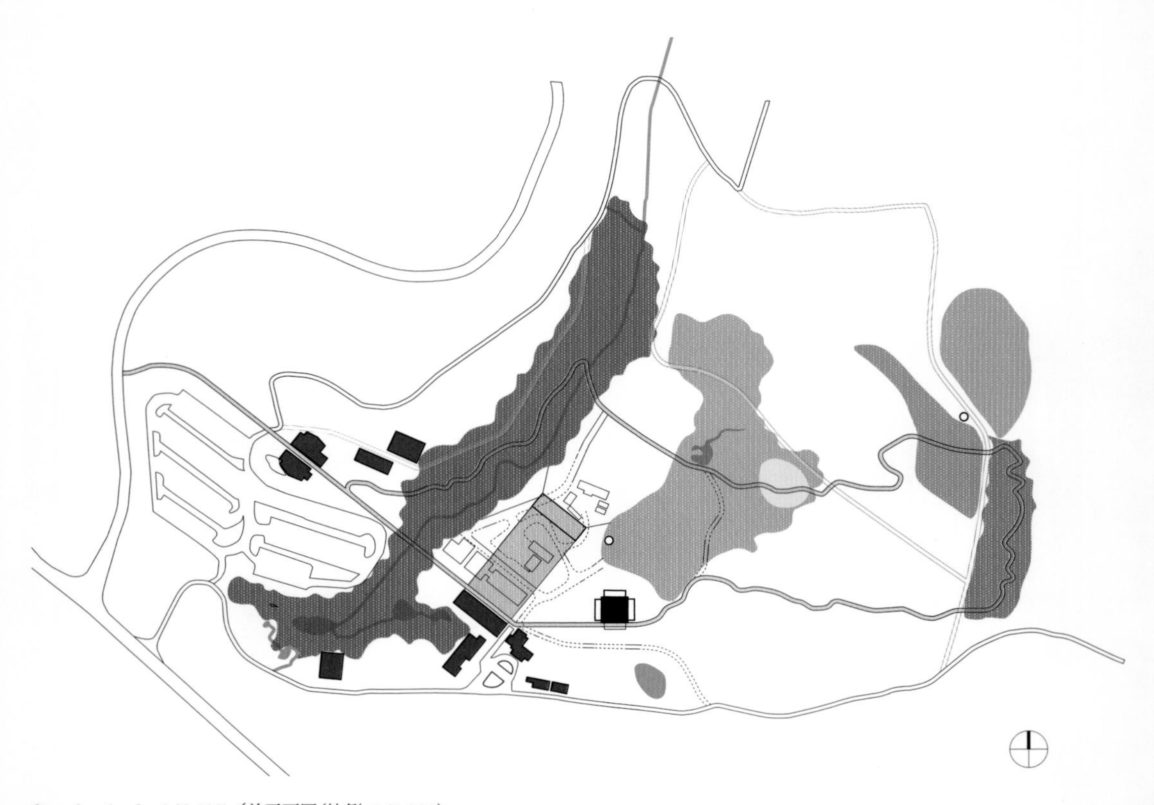

Site plan (scale: 1/5,000)／总平面图(比例: 1/5,000)

pp. 144–145: Exterior rendering of the conservatory. This page: Interior courtyard rendering. All images on pp. 144–151 by courtesy of the Architect.

第144-145页：温室外部效果图。本页：室内庭院效果图。

Ground floor plan (scale: 1/400)／一层平面图（比例: 1/400）

伊恩·波特国家温室位于澳大利亚首都堪培拉的澳大利亚国家植物园内。该植物园聚集了全国范围内的本土植物，具有重要意义；这里也被策划用于教育、植物研究和栽培。处在花园般的环境中，这个新的中心支持着稀有及濒危热带植物的展示、保护和繁殖工作，也将为植物园带来新的看点。

设计关注的核心在于：在这个气候变迁的时代，温室如何带给人独特且难忘的体验?这个问题在堪培拉的严冬环境中，得到了很好的解决。通常，从温室类型学的角度来看，使用轻型框架、全玻璃外壳会使室内在冬季因保温性差而过冷，在夏季则因过度暴晒而过热。因此，该项目提出了一种能使温室符合植物保护原则，并适应周围环境及气候变化特殊性的方法。

这个立方体仿佛悬停在景观之上，并人为框出一方天空。中央的空隙为热带植物的茁壮成长提供了必要的光照条件，同时限定了外部透明表皮的面积。围绕着这条空隙的层状表皮充当了高性能的隔热结构。表皮最外层利用太阳能来预热空气，并将多余的热量存储在辅助墙中；因而，在冬季它可以使建筑保持较高的室内温度，在夏季它可以使室内高温得到缓解。建筑的最外层根据其方位，从透明过渡到半透明，从而营造出独特的视觉效果。该建筑通过体量与光线的相互作用，营造出极具感染力的内部空间，并将世界领先的环保性能与高审美的空间体验结合在一起。

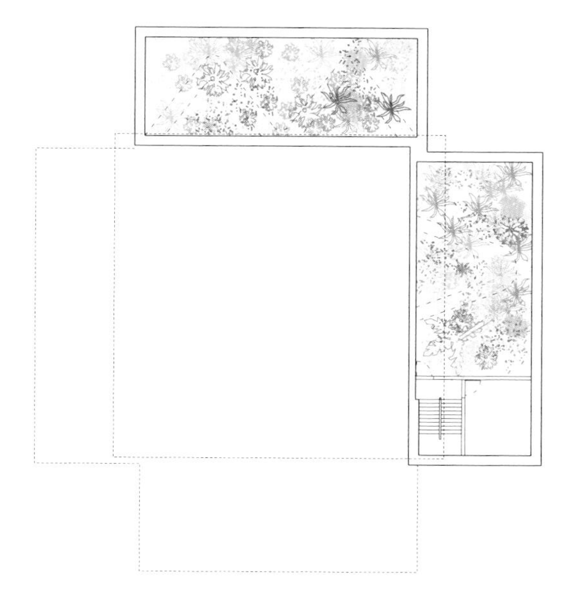

Basement floor plan／地下一层平面图

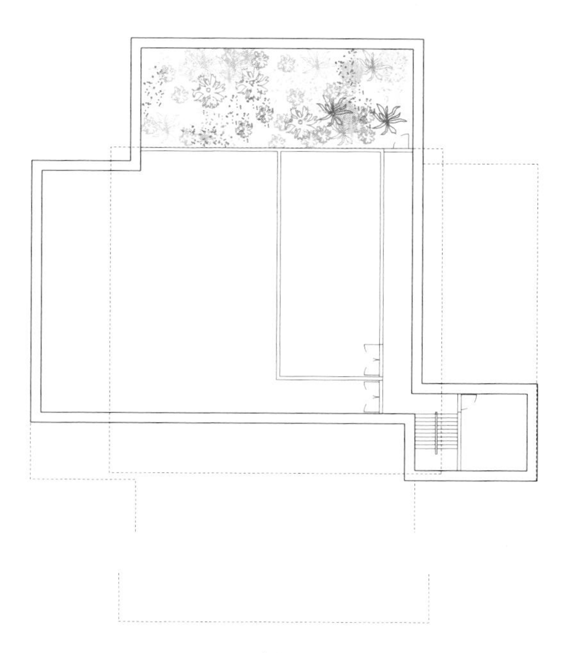

Second basement floor plan／地下二层平面图

SUMMER, DAYTIME

夏季，昼

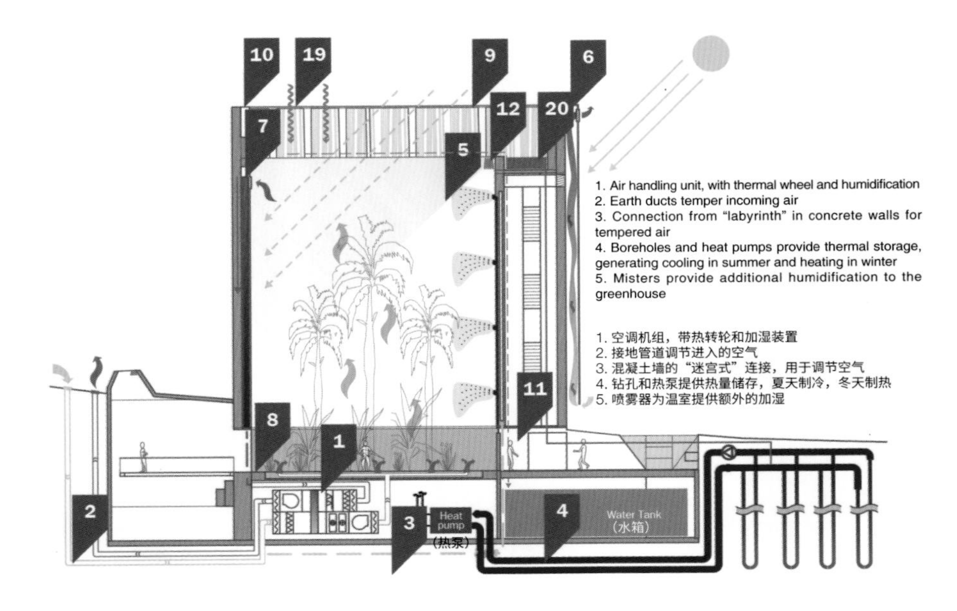

1. Air handling unit, with thermal wheel and humidification
2. Earth ducts temper incoming air
3. Connection from "labyrinth" in concrete walls for tempered air
4. Boreholes and heat pumps provide thermal storage, generating cooling in summer and heating in winter
5. Misters provide additional humidification to the greenhouse
1. 空调机组，带热转轮和加湿装置
2. 接地管道调节进入的空气
3. 混凝土墙的"迷宫式"连接，用于调节空气
4. 钻孔和热泵提供热量储存，夏天制冷，冬天制热
5. 喷雾器为温室提供额外的加湿
Heat pump
(热泵)
Water Tank
(水箱)

SUMMER, NIGHT TIME

夏季，夜

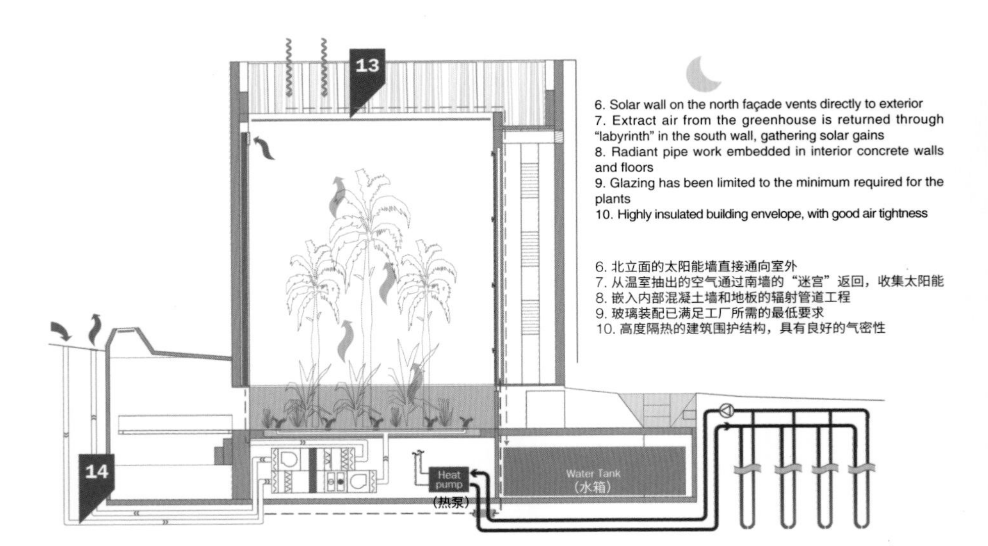

6. Solar wall on the north façade vents directly to exterior
7. Extract air from the greenhouse is returned through "labyrinth" in the south wall, gathering solar gains
8. Radiant pipe work embedded in interior concrete walls and floors
9. Glazing has been limited to the minimum required for the plants
10. Highly insulated building envelope, with good air tightness
6. 北立面的太阳能墙直接通向室外
7. 从温室抽出的空气通过南墙的"迷宫"返回，收集太阳能
8. 嵌入内部混凝土墙和地板的辐射管道工程
9. 玻璃装配已满足工厂所需的最低要求
10. 高度隔热的建筑围护结构，具有良好的气密性
Heat pump
(热泵)
Water Tank
(水箱)

Seasonal Environmental Diagram ／季节环境分析图

WINTER, DAYTIME

冬季，昼

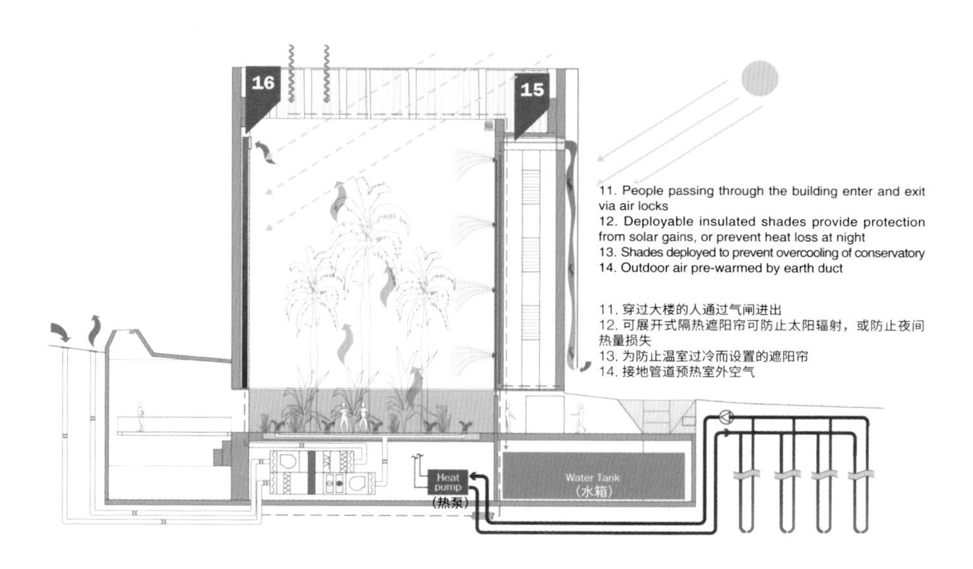

16
15
11. People passing through the building enter and exit via air locks
12. Deployable insulated shades provide protection from solar gains, or prevent heat loss at night
13. Shades deployed to prevent overcooling of conservatory
14. Outdoor air pre-warmed by earth duct
11. 穿过大楼的人通过气闸进出
12. 可展开式隔热遮阳帘可防止太阳辐射，或防止夜间热量损失
13. 为防止温室过冷而设置的遮阳帘
14. 接地管道预热室外空气
Heat pump
(热泵)
Water Tank
(水箱)

WINTER, NIGHT TIME

冬季，夜

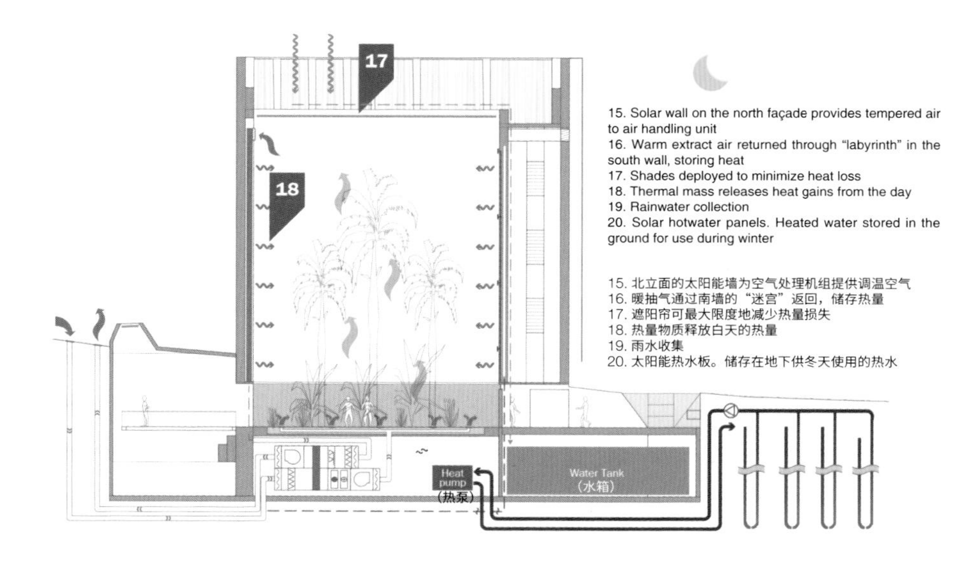

17
18
15. Solar wall on the north façade provides tempered air to air handling unit
16. Warm extract air returned through "labyrinth" in the south wall, storing heat
17. Shades deployed to minimize heat loss
18. Thermal mass releases heat gains from the day
19. Rainwater collection
20. Solar hotwater panels. Heated water stored in the ground for use during winter
15. 北立面的太阳能墙为空气处理机组提供调温空气
16. 暖抽气通过南墙的"迷宫"返回，储存热量
17. 遮阳帘可最大限度地减少热量损失
18. 热量物质释放白天的热量
19. 雨水收集
20. 太阳能热水板。储存在地下供冬天使用的热水
Heat pump
(热泵)
Water Tank
(水箱)

Essay:

The Vibe

Tom Heneghan

论文：

澳大利亚的“风气”

汤姆·赫尼根

In the classic Australian comedy movie *The Castle*, which I watched during my first visit to Australia, in 1997, an incompetent small-town lawyer argues in court – surprisingly successfully – that the government has no right to seize the land on which his client's family home stands in order to extend the runways of Sydney Airport. While the government's compulsory purchase of the land breaks no laws, the lawyer argues that it offends the Australian nation's prevailing "Vibe" – the sense of fairness to others on which Australians pride themselves, and the awareness that in our dealings with others there is always a "right thing to do" and a "wrong thing to do".

That "Vibe" seemed to me to explain the insistence of the nation's architects on their obligation to "do the right thing" about their shared physical environment. The raw beauty of the landscape and the power of its climate were respected and valued, and care was taken to minimize the damage that modern man's self-indulgent lifestyle was inflicting on our Spaceship Earth. It was taken as a given that the design of new buildings must be informed by the principles of passive design. In residential architecture, air-conditioners were few, and house temperatures were usually controlled by ingenious arrangements of louvred sun-shades, screens, and wise orientation. These lowenergy strategies were not the result of imposed government legislation – as they often were, for example, in Northern Europe. They were the results of the Australian people's consensus that passive architecture was "the right thing to do". They were also influenced, at least to some degree, by the example of the indigenous Australian community, whose teaching that buildings should "touch the earth lightly" has become an internationally influential slogan. For the Aboriginal people of Australia, the land is not spiritually inert, but is their still-living forebear – a proposition that can be well understood by those who visit and experience the indescribable majesty and vivacity of the sunset-glowing Uluru (formerly known as Ayers Rock). That particular Australian "Vibe" – the general consensus that architecture must respond ethically to the natural environment – was fascinating to me because, at that time most architects in Japan seemed to have little knowledge of, and indeed little interest in environmentally sustainable design. While Australian architects sought improved "passive" strategies that would minimize civilization's damage to the planet, climate issues in Japan were usually solved by the addition of more technologies and larger air-conditioners.

Eight years ago I was appointed one of the judges assessing entries for that year's Sydney-area Commercial Building of the Year award – a category which typically attracted large officetower projects. Of the ten judges, only two were architects, and we two sat huddled together, anticipating arguments that we felt we would be obliged to pursue with the hard-nosed business folk around the table – developers, estate agents, commercial lawyers, etc., who we assumed would prioritize the buildings' economic profitability, and have little interest in sensitive subjects such as the buildings' environmental performance. We were astonished, then, when the first speaker - an estate agent - insisted, to the universal agreement of all judges, that we should refuse to consider for the awards any building for which sustainability had not, clearly, been a design priority. Tenant companies, he argued, refused to rent high-energy-cost buildings, or those where the office staff could breathe only processed air. Sustainable design, all judges argued, "made economic sense".

1997年我第一次去澳大利亚时，看了澳大利亚的经典喜剧电影《城堡》，片中一个无能的小镇律师在法庭上胜诉，认为政府无权为了扩建悉尼机场的跑道而征用他委托人一家所住的土地。虽然政府强制征地的行为没有触犯法律，但律师指出，这触犯了澳大利亚人普遍认可的“Vibe”（风气、气氛），即他们引以为豪的公正感，以及与人交往时约定俗成的是非观。

在我看来，这种“风气”似乎可以解释，为什么澳大利亚建筑师坚持“正确对待”公共环境。他们敬畏、重视原始的景观之美和气候的力量，尽量减少现代人自我放纵的生活方式对地球这一“宇宙飞船”造成的损害。新的建筑必须遵循被动式设计的原则，已经被视为理所当然。空调很少在住宅建筑中出现，房屋温度常常由遮阳百叶、屏障和朝向来灵活调整。之所以应用这些低能耗策略，并不是像通常那样（例如在北欧）源于政府的硬性规定，而是因为澳大利亚人一致认同，被动式建筑是“对的事”。在某种程度上，这还受到了澳大利亚原住民社群的影响：他们主张的理念——建筑应“轻轻触摸大地”——已颇具国际影响力。对于原住民来说，这片土地不是死气沉沉，他们的先辈仍然活着，只要到访过夕阳下的乌鲁鲁（原名艾尔斯岩），感受那难以形容的雄伟与生机，就能很好地理解这一点。我着迷于澳大利亚这种“风气”，即人们普遍认为建筑应与自然环境相适应，而当时大多数的日本建筑师似乎都对环境的可持续设计知之甚少，甚至兴趣不大。在澳大利亚建筑师寻求更好的“被动式”策略以抑制人类文明破坏地球的时候，日本却常常在通过引进更多的技术和更大的空调设施来解决问题。

八年前，我曾作为“悉尼地区年度商业建筑奖”的评委之一，评审当年的参赛作品，它们大多是大型写字楼项目。十位评委中，只有两位是建筑师。我们俩挨坐着，猜测将不得不与桌边那些顽固的商业人士，例如开发商、地产代理、商业律师等，进行争论。在我们看来，他们应该会优先考虑建筑的经济效益，对环保性能等敏感问题则不感兴趣。因此，当第一位发言者——一位地产商——坚持认为应该拒绝考虑任何显然没有将可持续性作为设计重点的建筑，并且得到所有评审的一致认可时，我们目瞪口呆。他断言，租户公司会拒绝租用高能耗、职员只能呼吸经过处理的空气的建筑。所有评委都认为，可持续的设计“有经济价值”。

1993年，当我为熊本县农业研究中心草地畜产研究所——一所面向年轻农民的研究中心和培训机构——做设计工作时，我把《澳大利亚的原木建筑》(*Rude Timber Buildings in Australia*)[1]一书中的照片贴在书桌正对的墙上，作为灵感的来源。书中的134张照片展示了以羊舍为主的多样农场建筑，其中大部分的设计者和建造者都是农场工人，而非建筑师或专业的工程师。这些建筑简单厚重，有种惊人的雄伟感。然而，外观并不是农场主的首要考虑。他们只是优先选择附近的廉价材料，用最简单的技术连接木板和钢板，从而组装起这些棚舍。尽管如此，他们最终却实现了一种原始、质朴的美感。建筑撰稿人还试图指出，这些“粗野”的农场建筑与格伦·马库特的系列作品之间存在密切关联。不过两者仍有区别，最重要的一点就在于，马库特的作品毫无疑问是建筑，而不是纯粹的堆砌物。他设计的所有部件都经过深思熟虑，细节精准，绝不含糊。这些建筑也由此远离了可持续设计常给人留下的粗糙、低技的印象，而这种印象更多是源于建筑外观而非实质。马库特还指出，“可持续建筑往往并不是真的可持续——它甚至往往不是建筑！”

In 1993, when I was working on the design of the Kumamoto Grasslands Agricultural Institute – a research center and training school for young farmers – I stuck on the wall facing my desk, as sources of inspiration, photographs from the pages of the book *Rude Timber Buildings in Australia*[1] in which the 134 photographic plates illustrated a wide variety of farm buildings – mainly sheep-sheds – most of which had been designed and erected by farm-workers rather than by architects or professional constructors. The buildings had a simplicity and brute strength that gave them a dramatic grandeur. But, their appearance had not been a priority of their owners. The priority was that it be possible to assemble the sheds from materials that could be cheaply obtained nearby, using the easiest techniques for joining together planks of wood and sheets of steel. The resulting buildings, though, had a primitive, frugal beauty, and it has been tempting for architectural writers to suggest an affinity between these "rude" farm-buildings and the subsequent works of Glenn Murcutt. The critical difference between the two, though, is that in Murcutt's works there is no doubt – at all – that this is architecture rather than construction. All parts of Murcutt's buildings are carefully conceived and very precisely detailed. Nothing is approximate. That distances them very far from the raw, low-tech imagery that usually characterizes works of sustainable design, which are often based more in image than on substance. As Murcutt has pointed out, "sustainable architecture is often not very sustainable – and it's often not even architecture!"

The refinement of Murcutt's architecture has influenced the ambitions of contemporary Australian architects, and moved sustainable design on from its origins in small-scale residential works. We can now see how, in Murcutt's own recent works, and in the large-scale works of the other architects presented in this edition of *a+u*, how sustainable design has become the drivingforce of contemporary urbanism. Works like 1 Bligh Street (See pp. 156~167) and International House Sydney (See pp. 168~179) acknowledge the continuity of Australia's environmental "Vibe". They are works of very large scale, but their sustainability strategies are developed forms of those that Australian architects have been experimenting with for many decades. These buildings show how a city can be more than an efficient venue for commerce – it can and must offer visions of how we can live with each other, and with nature.

Note:

1. Philip Cox, John Freeland and Wesley Stacey (photographer), *Rude Timber Buildings in Australia* (Thames & Hudson, 1969).

马库特建筑的精妙性影响了当代澳大利亚建筑师的抱负，也推动着可持续设计从小型住宅项目开始不断向前发展。我们可以从马库特近期作品以及本书介绍的其他建筑师的大型作品中看到，可持续设计如何成为当代城市主义的驱动力。类似布莱街1号（见156~167页）、悉尼国际大厦（见168~179页）这样的作品，也印证了澳大利亚“风气”的延续。作为大规模的项目，它们的可持续策略都基于澳大利亚建筑师数十年的试验。这些建筑展示出城市不仅是高效的商业场所，还可以也必须提供一幅我们如何与他人共处、与自然共生的图景。

注释：

1. 菲利普·考克斯，约翰·弗里兰，韦斯利·斯泰西（摄影师），《澳大利亚的原木建筑》（泰晤士—哈得逊出版社，1969）

Tom Heneghan Chair of Architecture at the University of Sydney from 2001 to 2009. Heneghan is now a Professor of Architectural Design at Tokyo University of the Arts.

汤姆·赫尼根 2001 年至 2009 年任悉尼大学建筑系主任，现为东京艺术大学建筑科教授。

Ingenhoven Architects + Architectus

1 Bligh Street

Sydney, New South Wales 2011

英格霍芬建筑师事务所 + Architectus 建筑师事务所

布莱街 1 号

新南威尔士州，悉尼 2011

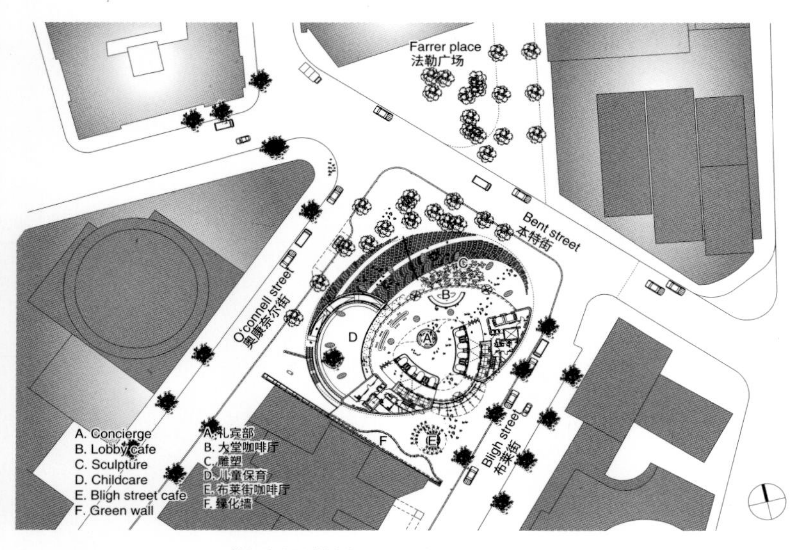

Site plan (scale: 1/2,000) ／总平面图（比例: 1/2,000）

TAXI
BRIDGE

Located in the historic heart of Sydney CBD, 1 Bligh Street sits within the valley created by the Macquarie Street ridge to the east and the tall buildings on George Street to the west. A shift in the street grid, together with a significant change in the topography, aligns the site diagonally with Sydney Harbor. The elliptical form amplifies this relationship, maximizing outlook while presenting an elegant, singular form when viewed from the harbor. The ground floor is conceived as a public room and offered to the city. Broad curving exterior steps, at the main entrance, rise to a sheltered public wintergarden, and provide a generous seating area engaging the immediate urban field.

The internal organization resists a centralized core to achieve visual expanse, inter-connectivity and communication. Close proximity of usable floor space to the facade is maximized, and contiguous column-free space creates flexibility. A naturally ventilated glass atrium extends the full building height. Glass elevators rapidly traverse the vertical section; they add to, and participate in, the dramatic spatial atmosphere. A doubleskin, glass facade operates as an environmental shield. The outer glass skin protects computer-controlled shades, shielding the double-glazed curtain wall from the sun, whilst reflecting natural light into the building. The strategy permits the use of low-iron glass, and achieves an ethereal lightness unexpected in the curtain wall tower skins. The atrium and breakout spaces to the south are naturally ventilated. Free heating is provided by in-slab pipework, supplied by the heat rejected via the cooling towers, and cooled by relief air from the office spaces. Passive strategies for environmental control are seamlessly integrated with emerging and highly efficient technologies, extending from air conditioning to a trigeneration system employing gas and solar energy to generate cooling, heating and electricity. Solar thermal collectors drive the cooling systems, waste water is treated and rainwater harvested to achieve efficacies throughout.

1 Bligh Street's Six-Star Green Star rating score is the highest awarded in New South Wales, and includes the maximum allowable five points for innovation in categories such as environmental design initiative, and for exceeding Green Star benchmarks. The tower culminates in a roof garden: a potent symbol of the project's ongoing contribution to the health of its occupants and the city.

p. 156: View of the tower along Sydney Harbour. p. 157: View of the tower from Loftus Street. Opposite, above: The open public ground floor of the tower invites people gather on the steps of the foyer. Opposite, below: Seating areas are provided on the atrium level of the public ground floor where occupants can have coffee or lunch. All images on pp. 156–167 by Hans Georg Esch.

第 156 页：沿悉尼港看到的该高层建筑外观。第 157 页：从洛夫图斯街看到的该高层建筑外观。对页，上：该高层开放的公共底层空间使人们聚集在门厅前的台阶上；对页，下：公共底层的中庭空间设有休息区，供用户享用咖啡或午餐。

布莱街 1 号位于悉尼中央商务区历史悠久的中心地带，地处谷地，东边是山脊般的麦觉理街，西边是高楼林立的乔治街。路网的改变以及地形的显著变化使场地与悉尼港呈对角线状。椭圆的建筑形态强化了这一关系，使得面向海港的视野最大化，并且向身处海港的观者呈现出优雅、非凡的外观。底层作为公共空间，面向城市开放。主入口外部宽阔的曲线台阶通向一个有遮蔽的公共冬季花园，并提供了宽敞的休息区，与周围的城市环境融为一体。

内部的空间组织排除了采用集中式核心筒的设计，从而实现视野扩展、空间互联以及用户交流。可用地面空间与外立面之间的距离被无限拉近，连续的无柱空间创造了灵活性。自然通风的玻璃中庭贯通整个建筑。玻璃电梯在垂直空间快速穿梭，令人印象深刻，并成为塑造空间氛围的要素之一。双层玻璃幕墙犹如一道环境屏障。外层玻璃保护着电脑控制的遮阳板，使里层的双层玻璃幕墙免受阳光直射，同时将自然光反射到建筑内部。这一策略让低铁玻璃（也称超白玻璃、高透明玻璃）的使用成为可能，并在高层幕墙中实现了超出预期的轻盈感。中庭和南面的休息空间自然通风。冷却塔排出的热量可通过楼板内的管道系统为内部空间免费供暖，并通过办公空间释放的冷气得到冷却。被动式环境调控策略与新兴的高效技术无缝结合，其范围之广，覆盖了从空调到利用天然气和太阳能制冷、供暖和发电的三联产系统。太阳能集热器驱动制冷系统，废水被处理，雨水被收集，以此实现所有环节的高效利用。

布莱街 1 号获评“绿色之星”（Green Star，澳洲绿色建筑星级评价体系）六星级，这是新南威尔士州最高奖项，并因其在环境设计新提案门类中，超过了绿色之星的标准而获得满分 5 分。这座高层建筑还在顶层设置了屋顶花园，这也是该项目持续为用户健康和城市运转做出贡献的有力表现。

Opposite: Workspace balconies facing the atrium and glass lift cars enhance vertical communication among its tenant.

对页：办公区的阳台面向中庭，玻璃的电梯可增强用户间的垂直交流。

13

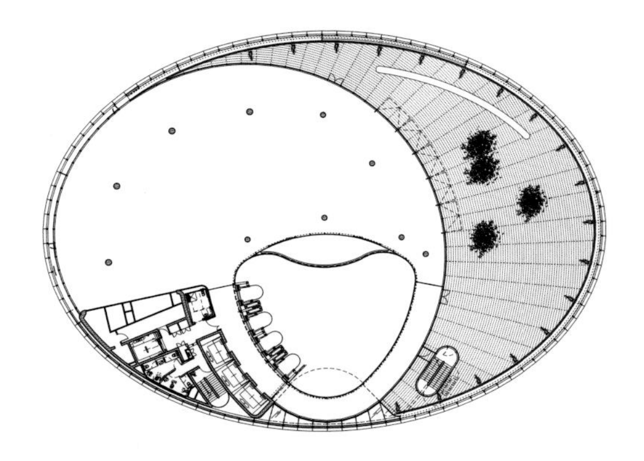

28th floor plan／二十八层平面图

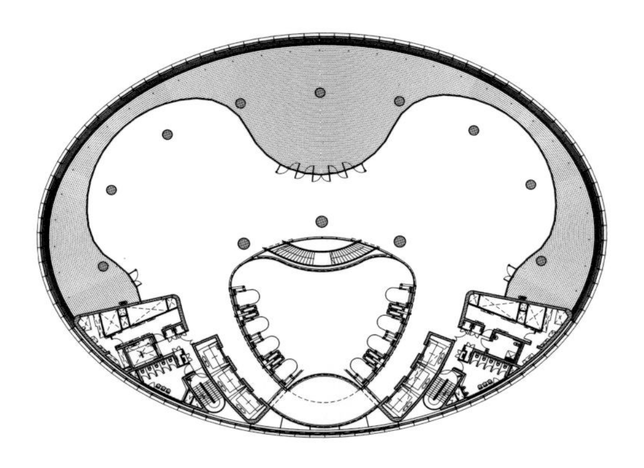

15th floor plan／十五层平面图

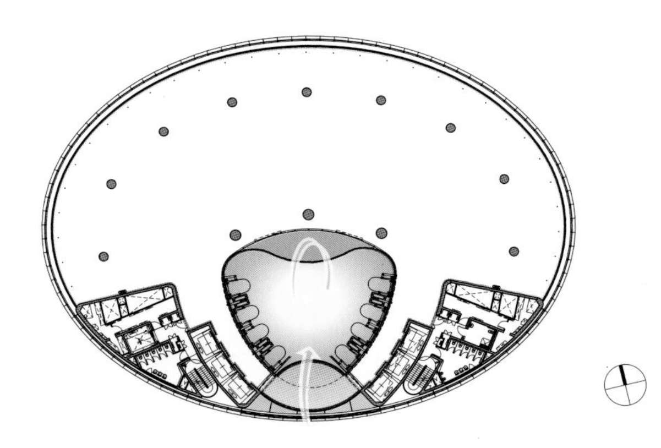

Low rise floor plan (scale: 1/1,000)／低层平面图(比例: 1/1,000)

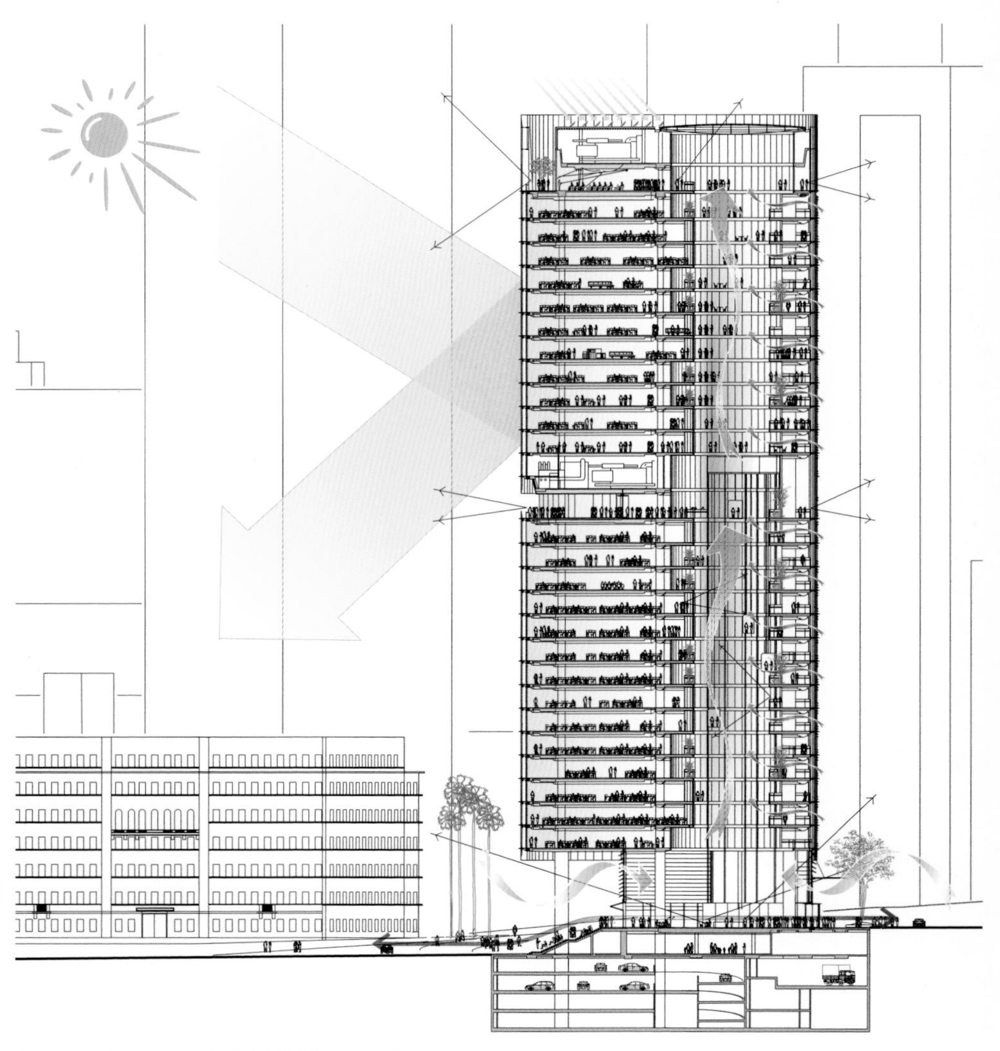

Section (scale: 1/1,500)／剖面图（比例: 1/1,500）

pp. 162–163: View from the atrium looking up. The atrium is 120m in height and kept naturally ventilated. p.165: Aerial view of the tower from the south. With the solar PV cells located on the highest part of the roof, the roof garden located on a lower level to provide its tenants with a comfortable social environment.

第 162-163 页：从中庭仰望建筑。中庭高度为 120 米，并保持自然通风。第 165 页：从南侧俯瞰该高层建筑。太阳能光伏板位于楼顶最高处，屋顶花园稍低于它，为其租户提供舒适的社交环境。

Credits and Data

Project title: 1 Bligh Street
Developer: Dexus Property Group, Dexus Wholesale Property Fund, Cbus Property
Location: 1 Bligh Street, Sydney, New South Wales, Australia
Completion: July 2011
Architect: Ingenhoven Architects + Architectus
Design Team (Architectus): Ray Brown (Managing Director), Mark Curzon, Simon Zou, Linda Bennett, Scott Hunter, Ryan Townsend, Daniela Salhani, Nikhil Fegade, Karolin Baer, Tommy Ford, Fawzi Soliman, Michael Harrison, Stewart Verity, Murray Donaldson, Camille Lattouf, Rodd Perey, Ryan Hanlen, Harry Broekhus, David Kamel, Darrin Rodrigues, Chase Ronge, Siera Chuah, Annette Gall
Design Team (Ingenhoven Architects): Christoph Ingenhoven (Principal), Martin Reuter (Director), Christian Kawe, Martin Slawik, Thomas Weber, André Barton, Mario Böttger, Elisabeth Broermann, Darko Cvetuljski, Ralf Dorsch-Rüter, Hye Jin Jung, Christian Kob, Andrea König, Alice Koschitzki, Dr. Mario Reale, Evelyn Scharrenbroich, Ulrike Schmälter, Alexander Schmitz, Jürgen Schreyer, Brett Stover, Erich Tomasella, Lutz Büsing, Felix Winter
Consultants: APP (Project Manager), Enstruct (Structural Engineer), Enstruct/ ARUP (Structural Steel), Cundall (ESD), DS Plan AG (Germany) / Arup / Enstruct (Façade), ARUP (M&E Services, Acoustics), Morris Goding Accessibility (Accessibility), Masson Wilson Twinney (Traffic), Steve Paul & Partners (Hydraulic, Fire), Tropp Lighting Design, Arup (Lighting), Sue Barnsley Design (Landscape), MEL Consultants (Wind Engineer), NDY (Vertical Transport), Exparrot (Green Wall), Design 5 – Architects (Heritage Consultant), Davis Langdon (BCA + PCA)
Gross floor area: 45,760 m²
Project estimate: $285 million AUD

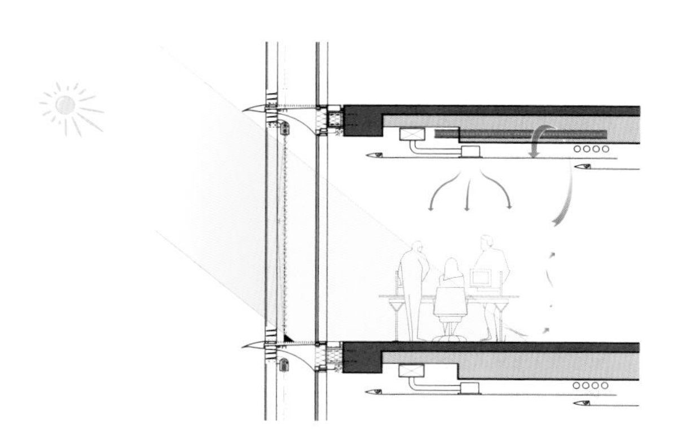

Double skin facade detail – solar protection and cooling diagram

双层幕墙细节—日照防护及降温分析图

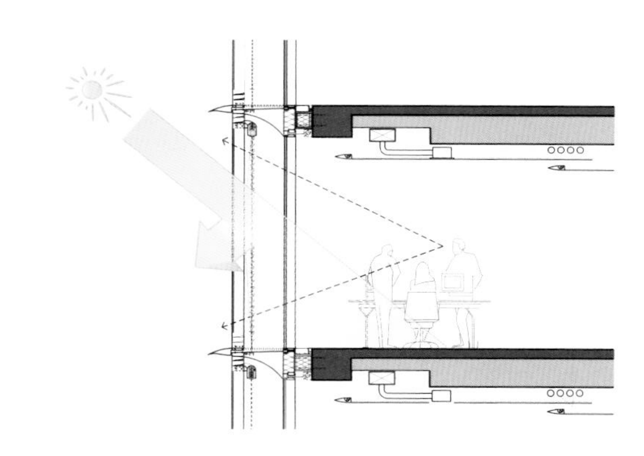

Double skin facade detail – solar protection and maximum views

双层幕墙细节—日照防护及视野最大化分析图

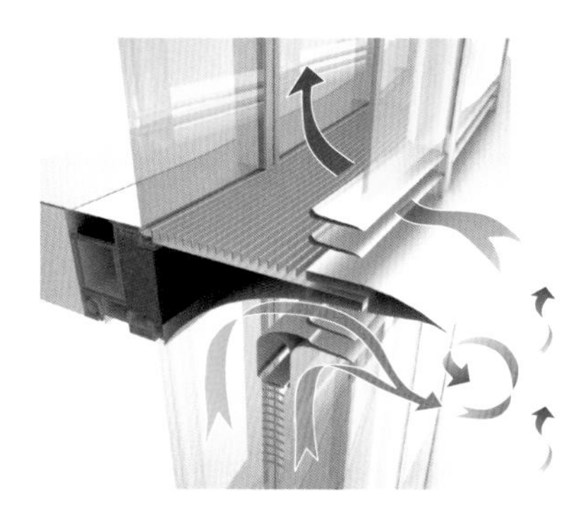

Double skin facade detail

双层幕墙细节

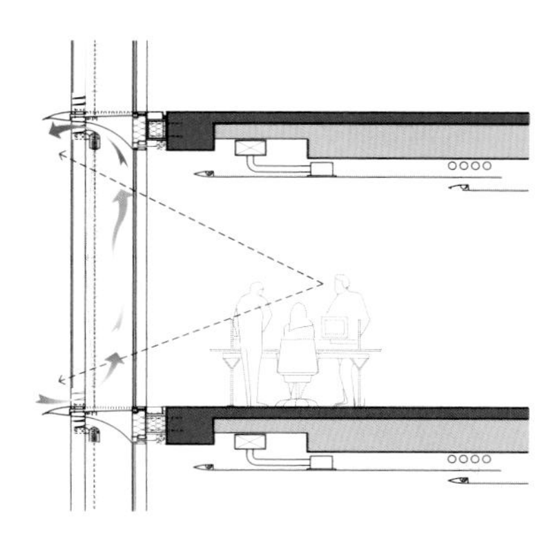

Double skin facade detail – air flow diagram

双层幕墙细节—气流分析图

Tzannes

International House Sydney

Barangaroo, New South Wales 2017

哲纳司建筑师事务所

悉尼国际大厦

新南威尔士州，巴兰加鲁 2017

International House is located in Barangaroo,a newly-developed precinct at the western edge of Sydney's CBD. Since 2012 the area has been transformed from a disused container terminal to a 22-hectare waterfront development for commerce, housing, recreation and leisure. This ongoing development uses a centralized system for recycling, renewable energy regeneration, and embedded electricity, and deploys Sydney Harbor water for cooling. It will advance knowledge on the design and delivery of low carbon precincts. International House addresses the city and connects Barangaroo with the historic heart of the commercial district of Sydney.

The project is currently the world's tallest commercial timber building. Six levels above the public domain are constructed entirely from engineered timber, including floors, columns, walls, roof, lift shafts, egress stairs and bracing bays. Using 3,500-m^3 of sustainably-grown and recycled timber, the structure is comprised of cross-laminated timber (CLT), laminated veneer lumber (LVL), Glulam technology comprising recycled Australian hardwoods. With a low carbon footprint and close to zero waste in the construction process, the system provides a sustainable alternative to conventional construction practices, and off-site prefabrication ensures efficiency and quality control in construction. The building demonstrates the commercial viability of mass timber construction, and thus increases the opportunity for architecture to effectively contribute to a sustainable future.

A legible and homogenous architectural framework defines the interior; infused with the texture and smell of pine, it provides a rich range of sensory experiences. At street level, a doubleheight colonnade in recycled ironbark and turpentine timbers distinguishes the architecture, and both shapes and elevates the public domain. Encased in low-iron glass, the clarity and lightness of the structural system is revealed, allowing an appreciation of the architecture in its totality.

The restrained logic and uncompromised environmental credentials of the laminated timber construction informs an architectural aesthetic that is both raw and elegantly refined.

accenture

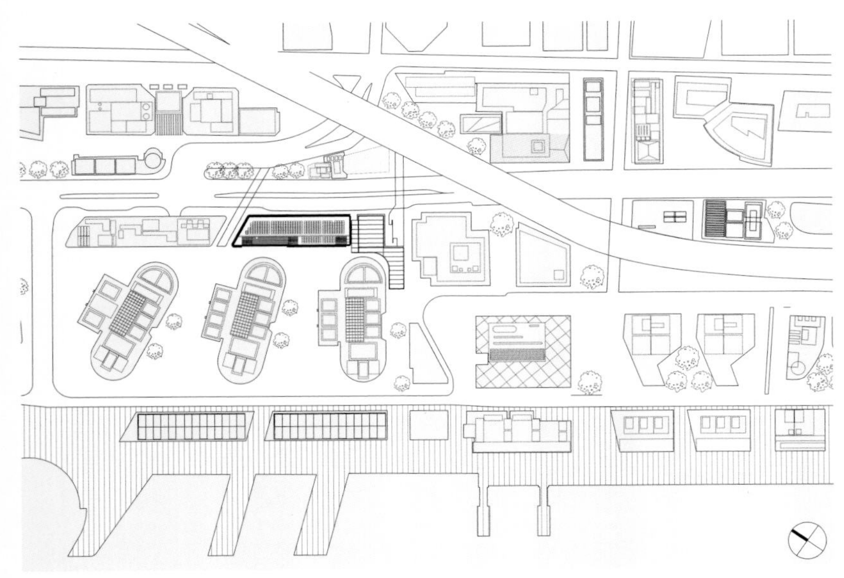

Site plan (scale: 1/5,000) ／总平面图（比例: 1/5,000）

p. 169: The building sits on the eastern edge of Barangaroo South. This page: View of the 6-storey building from Sussex street. All images on pp. 168–179 by Ben Guthrie.

第 169 页：建筑位于南巴兰加鲁的最东边。本页：从苏塞克斯大街看向这栋六层高的建筑。

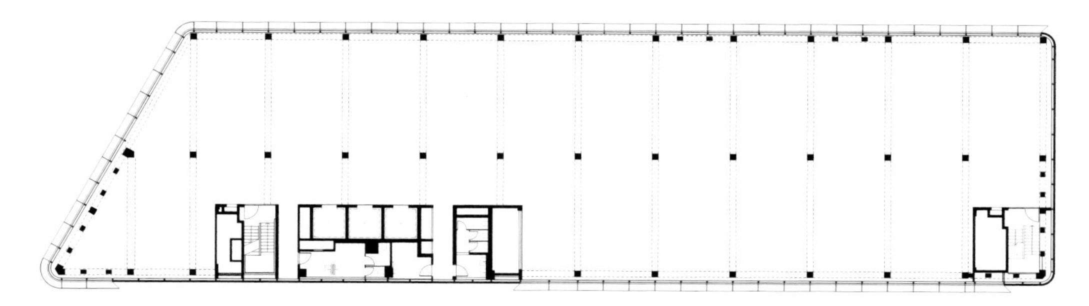

Typical floor plan／标准层平面图

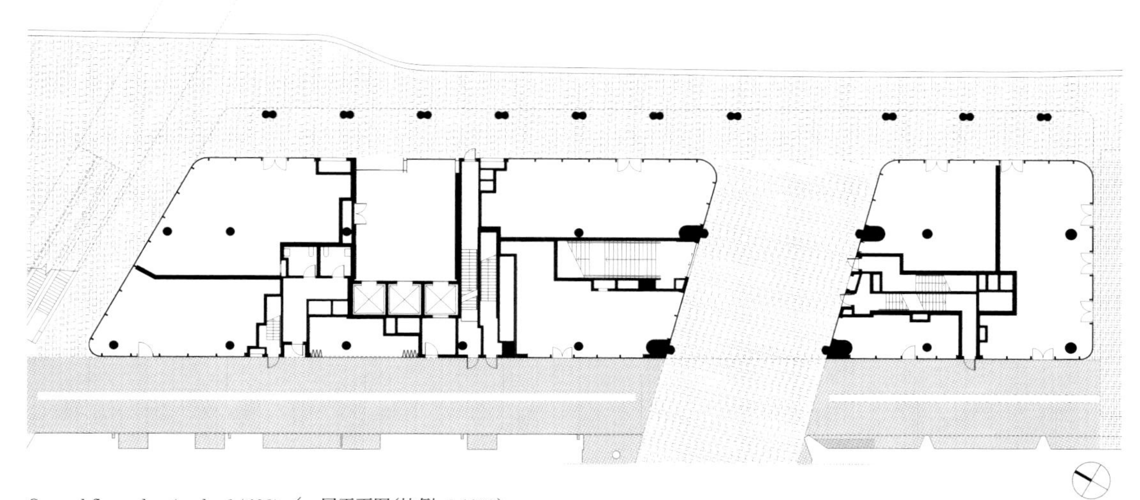

Ground floor plan (scale: 1/600)／一层平面图(比例: 1/600)

Credits and Data

Project title: International House Sydney
Client: Lendlease
Location: 3 Sussex Street, Barangaroo, New South Wales, Australia
Completion: April 2017
Architect: Tzannes Team
Design Team: Jonathan Evans (Design Director), Alec Tzannes (Design Director), Dijana Tasevska (Project Architect), Tony Lam (BIM and Documentation Lead), Chi Melhem, Lily Tandeani, Amanda Roberts, Dustin Cashmore, Wenxi Ren, Carl Holder, Linda Kennedy
Consultants: Lendlease DesignMake (Timber Structural Engineer), Concrete Arcadis (Structural Engineer), AECOM (Head Services Consultant, Mechanical Engineer, Electrical, Lighting + Comms), Warren Smith & Partners (Hydraulic and Fire Services), Defire (Fire Engineering), ESD (Lendlease), Wilkinson Murry (Acoustic Engineer), Surface Design (Façade Engineer)
Construction Team: Lendlease Building (Principal Builder)
Gross Floor Area: approx. 8,000 m²

Opposite: The two-storey covered plaza provides connectivity for pedestrians from Mercantile Walk. This page: View along the double-height colonnade evocating an image of being in a forest of trees.

对页：两层高的带顶棚广场连通了商业步行街。本页：沿两层通高的柱廊望去，仿佛置身于森林之中。

悉尼国际大厦位于巴兰加鲁，该地是悉尼中心商务区最西边一片新发展起来的区域。自 2012 年以来，该区域已从废弃的集装箱码头转变为占地 22 公顷的滨水开发区，集商业、住宅、娱乐和休闲于一体。这个处于建设中的开发区采用了集中式回收系统、可再生能源系统及嵌入式电力系统，同时利用悉尼港的海水提供制冷。它将提升人们对于低碳区设计与使用的认知。悉尼国际大厦面向城市，并连接起了巴兰加鲁和悉尼商业区的历史中心。

该项目是迄今为止世界上最高的木质商业建筑。地上 6 层，包括楼板、柱子、墙体、屋顶、电梯井、逃生楼梯和结构斜撑，全部由工程木材构筑而成。该结构使用了 3,500 立方米可持续种植和循环利用的木材，其中包括正交胶合木、单板层积材和胶合技术压制的澳大利亚可循环利用硬质木材。整个过程实现了低碳施工，且几乎无废料产出，为传统施工提供了一个可持续的替代方案。除此之外，场外预制为施工效率和施工质量提供了保障。该建筑展示了大规模木制建造在商业项目中的可行性，从而增加了建筑为未来可持续发展作出有效贡献的机会。

清晰而同质的建筑框架定义了室内空间；整个建筑充满了松木的质感和气息，提供了丰富的感官体验。两层通高的临街柱廊由回收再利用的桉木和松脂木构筑而成，让人眼前一亮，它塑造并提升了整个公共区域的品质。建筑使用超白玻璃，露出清晰且轻盈的结构系统，使人们可以透过玻璃欣赏到整座建筑。

层压木结构有节制的逻辑，以及不折不扣的环保性能，共同形成了原始且优雅的建筑美学。

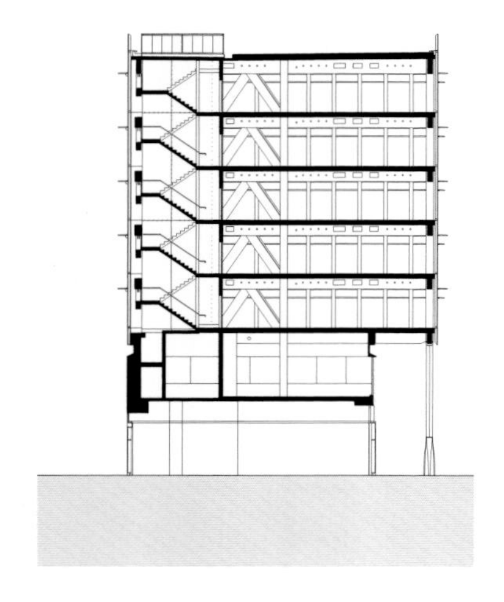

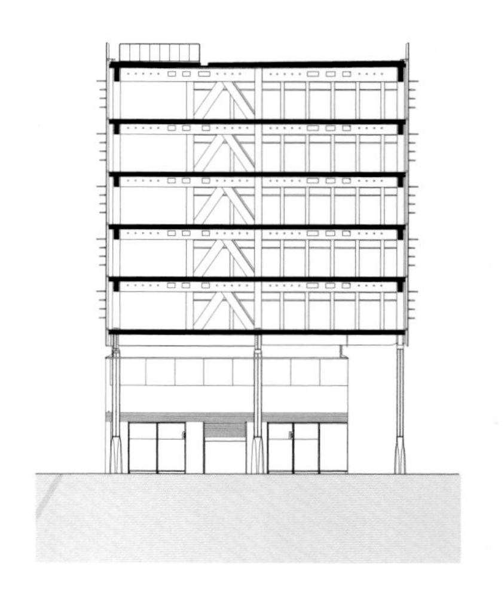

Section (scale: 1/600) ／剖面图（比例: 1/600）

Opposite: The colonnade along Sussex street is held up by angled recycled ironbark props secured on concrete bases.

对页：沿苏塞克斯大街的柱廊，由固定在混凝土基座上倾斜的支柱支撑，支柱由可循环使用的澳洲橡木制成。

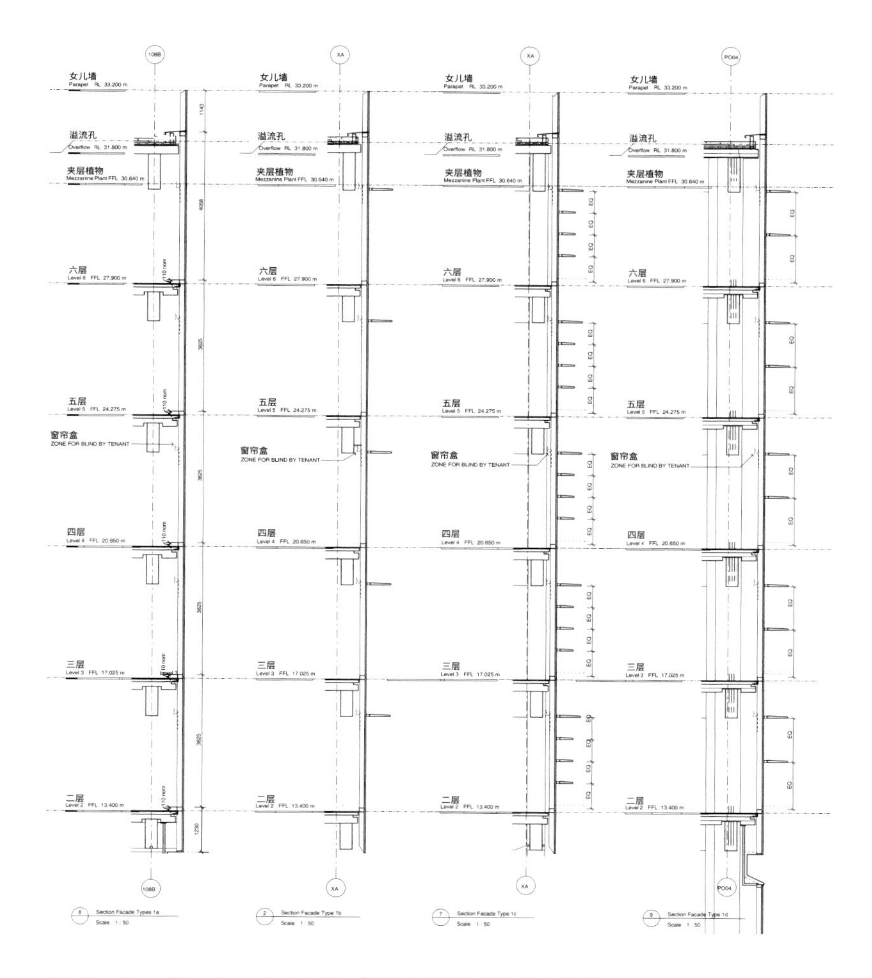

Detail section of facade (scale: 1/200) ／外立面剖面详图(比例:1/200)

Opposite, above: Cross-laminated timber panels form the floors and European spruce are used on the internal structure of the building. Opposite, below: General view of the interior where chilled beams, pipes and other building services are painted black and left exposed. p. 178: Encased in low-iron glass, the building's timber structural system is exposed, allowing it to be appreciated by anyone passing by. p. 179,above: View of the fire stairs, enclosed with fire-rated glass to provide natural daylighting and views to encourage stairs circulation between floors. p. 179, below: The texture and smell of pine in the interior provides a rich range of sensory experiences.

对页，上：楼板由交叉层压木板材构筑，室内结构则使用欧洲云杉；对页，下：室内冷梁、管道和其他建筑设备被涂成黑色并裸露在外。第178页：超白玻璃的使用使建筑的木质结构系统暴露在外，供过往行人欣赏。第179页，上：消防楼梯用防火玻璃封闭，提供自然采光和景观，并希望以此促进人们对楼梯的使用；第179页，下：松木的质地和气味为室内空间提供了丰富的感官体验。

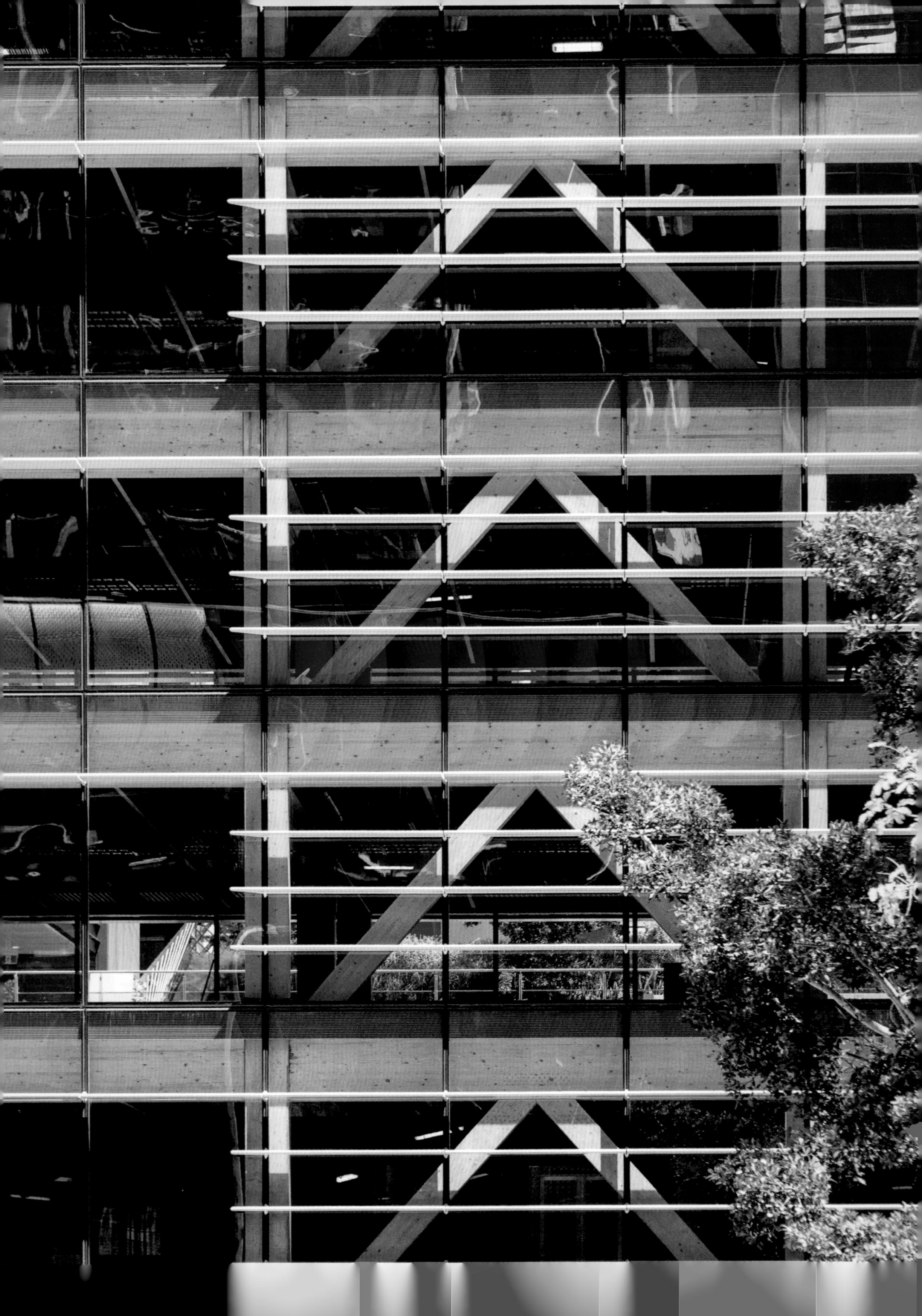

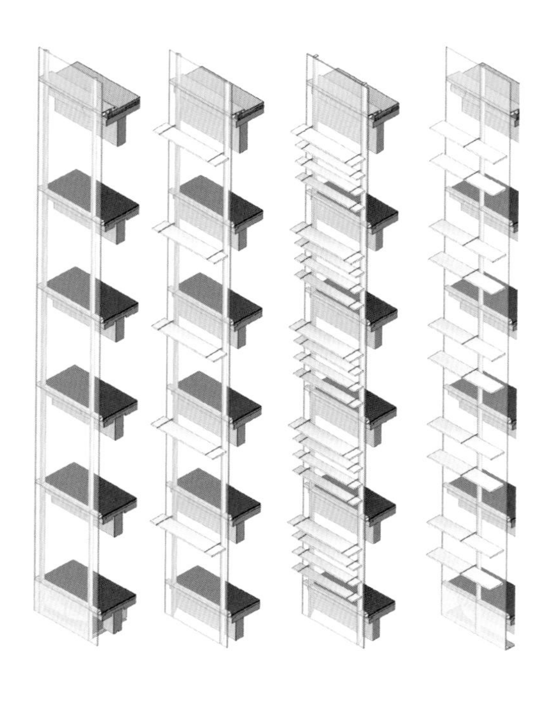

Facade detail／外立面细节详图

Tzannes

The Brewery Yard, Central Park

Chippendale, New South Wales 2014

哲纳司建筑师事务所

中央公园啤酒工业园

新南威尔士州，奇彭代尔 2014

LIVE IN CENTRAL PARK
EMERGENCY EVACUATION POINT
SALES

The Brewery Yard is an adaptive reuse of the former Irving Street Brewery in Central Park, Sydney. It transforms an important early 20th-century factory building to a power, heating and cooling plant for the entire Central Park precinct and allows for future adaptation and commercial use. Located on the park's western boundary, it defines the central green and serves as an iconic marker within the precinct.

The original building is valued as an architectural remnant and a signifier of the social, economic and architectural history of the site, and is protected by heritage legislation. Alterations and additions to the building were strenuously controlled. The project mediates the past and present through the juxtaposition of the industrial masonry architecture, with distinctive new elements and industrial systems. A major gas-powered, trigeneration plant is located under a newly-formed urban courtyard, and is connected to new infrastructure over the existing roof. It provides electricity and hot and cold water to approximately 255,000-m² of development, comprised of 2,200 residential apartments, 20,000-m² of retail space, significant public space – including 6,400-m² of public parkland – and commercial space. The use of this technology across the Central Park precinct has been modeled to predict a saving of 190,000 tons of carbon emissions over the 25-year life of the equipment. Recently the precinctwide trigeneration plant has been extended to power the adjacent university building infrastructure.

A zinc mesh wraps the new technology like fabric draped over a curved frame. It simultaneously refers to the orthogonal geometry of the masonry form below, and the curvilinear shape of the cooling tower plant within. The mesh transparency has been minimized to achieve formal legibility, whilst maintaining the required air intake. The historic brick chimney has been restored and lined with a new stainless steel inner tube to exhaust fumes from the subterranean plant. Other historic elements, such as brewing silos and hoppers, are retained in situ and are visible from the glazed northern wall. The careful balance of the demolished, the conserved and the explicitly new elements in this work establishes a bold and elevated standard for heritage conservation in Australia. Through the mediation of the highly technical program and its masonry container, typically invisible environmental infrastructure is given local material expression.

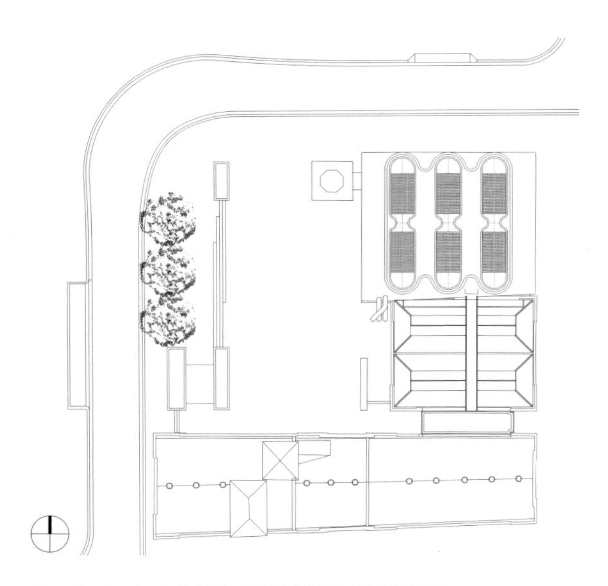

Site plan (scale: 1/1,000) ／总平面图(比例: 1/1,000)

pp. 180–181: Three zinc-cloaked towers form a power plant set atop the roof of the old beer brewery's boiler room. Image by Chris Jones. Opposite: Aerial view of site. The towers are the first stage of the scheme which in the later stages will include shops, a hotel and student housing and a public park. All images on pp. 182–189 by John Gollings unless otherwise noted.

第 180-181 页：三座镀锌塔在老啤酒厂锅炉房的屋顶上形成了一座发电厂。本页：场地鸟瞰图。塔楼是该方案的第一阶段设计，后期还将包含商铺、酒店、学生公寓和公园的设计。

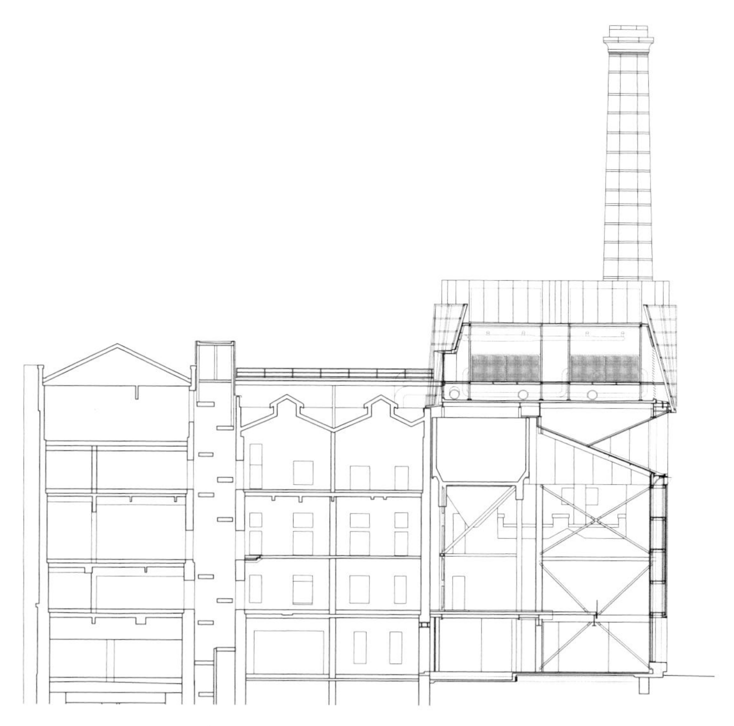

Section (Scale:1/600) ／剖面图(比例: 1/600)

啤酒工业园位于悉尼中央公园，是前欧文街啤酒厂的更新项目。这个项目将一个 20 世纪早期的重要工厂改造成一处能为整个中央公园区域供电、供暖和制冷的设施；设计师也考虑到了未来再改造和用作商业用途的可能性。它坐落于公园的最西侧，加深了这片中央绿地的轮廓，也是该区域的标志性建筑。

场地上原有的建筑被视为历史遗迹，这象征着这片区域的社会、经济和建筑历史，受遗产法保护。建筑的改造和扩建也因此受到了严格的控制。该项目通过使工业砌筑建筑与独特的新建筑元素及工业体系融合，来实现新与旧的调和。项目包含了 2,200 套住宅公寓、2 万平方米零售空间、包含 6,400 平方米公园的有效公共空间以及商业空间。一个大型燃气三联发电机位于新建的中庭下方，并与置于原屋顶结构之上的新设备相连，为整个项目约 25.5 万平方米的区域提供电力和冷热水供应。这项科技在整个中央公园的应用经过了预先的建模计算，预计设备在未来 25 年内将减少 19 万吨碳排放。近期，整个区域内的三联发电机的电力输出已得到了拓展，为邻近大学建筑的基础设施提供动力。

新的设备被镀锌网状面板包裹着，布料般从曲线框架上垂下来。它同时反映出下方砌体的正交几何形状，以及其内部冷却塔的曲线形状。建筑师通过将网状面板通透性最小化来获得屋顶形态的辨识度，同时保障冷却塔所需的进气。历史悠久的砖烟囱已被修复，内衬为全新的不锈钢内管，以排出来自地下设备的烟雾。与此同时，透过北侧的玻璃墙可以看到其他具有历史意义的元素，例如之前的酿酒仓和料斗，它们依然保留在原位。拆除的、保留的和新建的元素在这个项目中达成了平衡，也为澳大利亚遗产保护建立了一个大胆而高水准的范例。在高科技方案与砖石建筑容器的协调中，通常被隐藏起来的环境基础设施通过当地材料的使用被赋予了新表达。

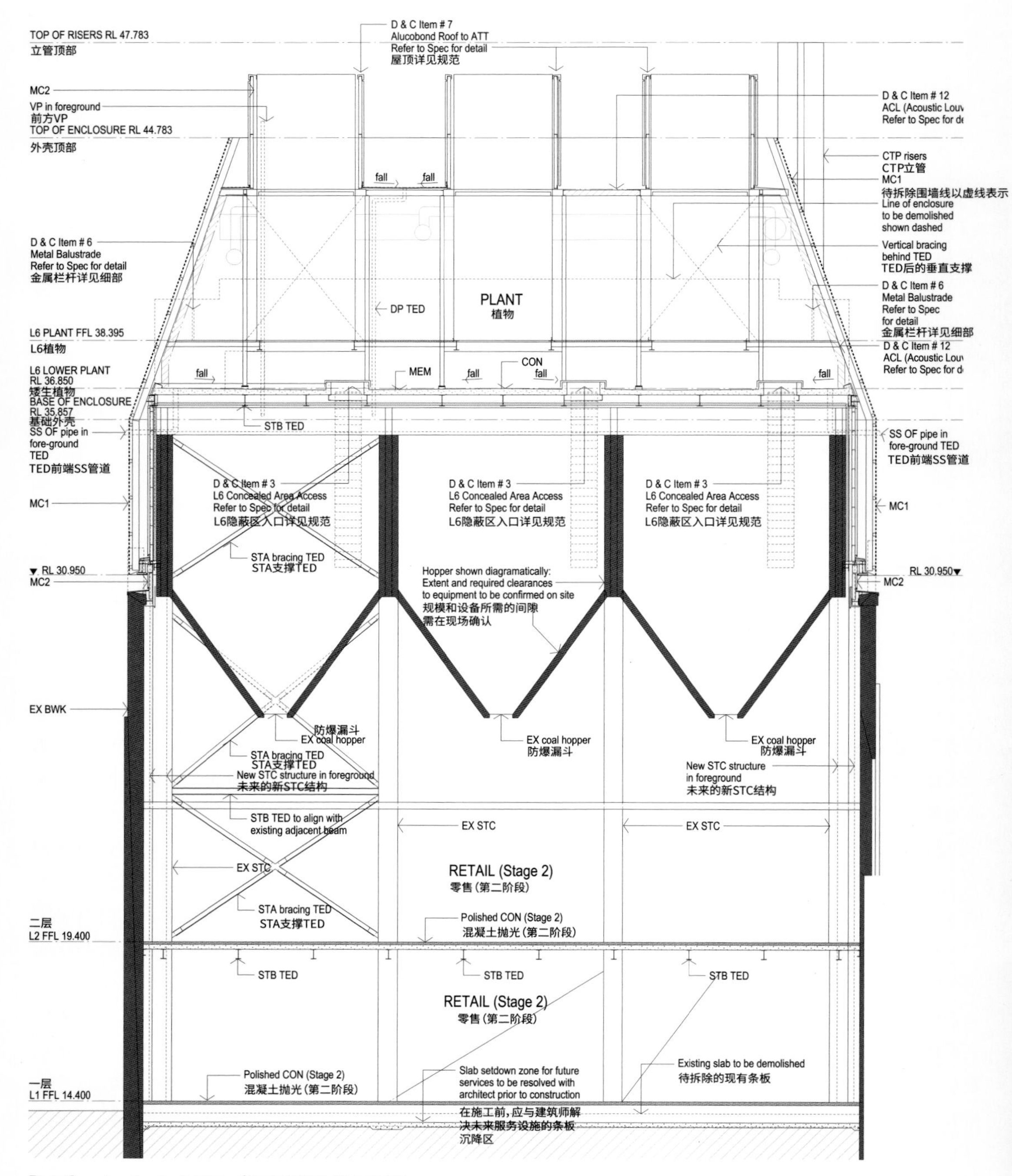

Detail section (scale: 1/200) ／剖面详图(比例: 1/200)

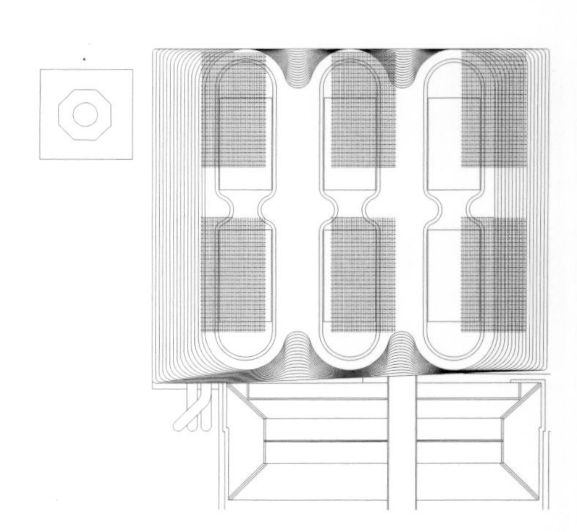

Roof plan／屋顶平面图

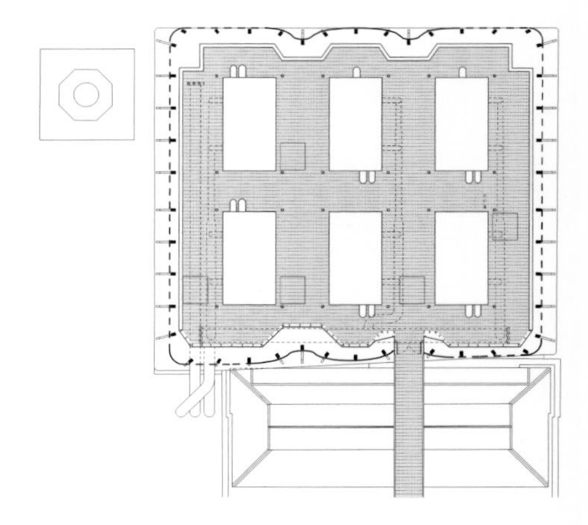

Upper plant level plan／上层工厂平面图

Credits and Data

Project title: The Brewery Yard, Central Park
Developer: Frasers Property Australia and Sekisui House Australia
Location: 5 Central Park Avenue, Chippendale, New South Wales, Australia
Completion: June 2014
Architect: Tzannes
Design Team: Allison Cronin (Project architect), Alec Tzannes (Design director), Amanda Roberts, Antoinette Cano, Ben Green, Bruce Chadlowe, Carl Holder, Derek Chin, Nadia Zhao
Consultants: Meinhardt (Structural Engineer), WSP Group (Electrical Engineer), WSP Group (Mechanical, Hydraulic, Fire Engineer), Webb Australia (Lighting), Acoustic Logic (Acoustic), WSP Built Ecology (Environmental), Atlus Page Kirkland (Cost consultant), Accessibility Solutions (Accessibility Consultant), City Plan Services (BCA Consultant), JBA Urban Planning Pty Ltd (Planner), Degotardi Smith & Partners (Surveyor), Urbis Heritage Consultant (Heritage and Conservation Consultant)
Construction Team: Christie Civil (Builder, Shroud), Total Constructions (Builder, Trigen Plant Equipment), Scott Clohessy (Project Manager)
Project area: 586 m²

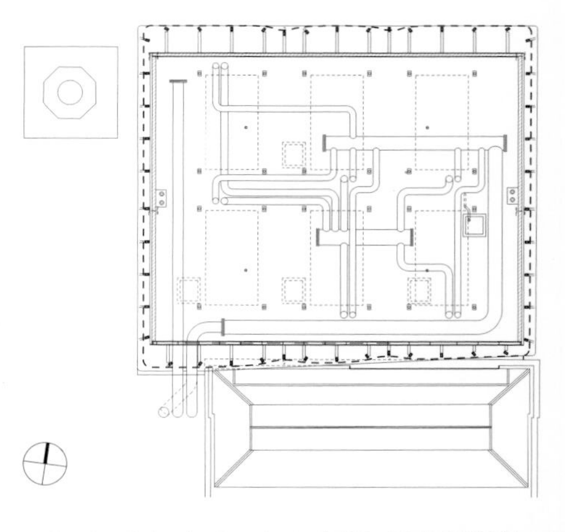

Lower plant level plan (scale: 1/500)／下层工厂平面图(比例: 1/500)

p. 184: Towers supported by steel beams forming a scaffolding-like structure is left exposed from one side of the building, reinforcing the former industrial use of the site. p.187: The red brick boiler room is retained, while the new steel work supporting the towers are left visible expressing the adaptive re-use of the building. Opposite: The metallic trigeneration towers are used to heat and cool water and provide electricity. This page, above: Zinc-meshed sheets are draped over the form like a piece of fabric. This page, below: Perforation of the mesh is kept minimum to retain the solidity of the overall form, while providing permeability for the cooling towers.

第184页: 由钢梁支撑的塔楼形成脚手架般的结构，从建筑物的一侧凸显出来，显示出该场地以前的工业用途。第187页: 红砖锅炉房被保留下来，而支撑塔楼的新钢架结构则外露，以表现该建筑的自适应性再利用。对页：金属三联发电机塔用来提供电力和冷热水。本页，上：镀锌网状面板，布料般从曲线的框架上垂下来；本页，下：最小化网状面板的孔洞以保证整体建筑形态的稳固性，同时保证冷却塔所需的透气性。

Atelier Jean Nouvel (Design Architect)
PTW Architects (Local-Collab Architect)

One Central Park

Chippendale, New South Wales 2014

让・努维尔建筑师事务所（设计方）
PTW 建筑师事务所（当地合作方）
中央公园一号
新南威尔士州，奇彭代尔 2014

One Central Park is a landmark development within the Central Park precinct. From Central Park's early inception, the expression of sustainable design — in the form of green open spaces and planting — was fundamental. The design and integration of the green walls within this complex residential tower has embraced, and continues to reaffirm, this early design intent.

The building comprises the 34-story East Tower and the 18-story West Tower, both held above a five-story podium that directly addresses a major city artery. The two towers hold 624 apartments, diverse in scale and layout. The podium contains retail, commercial and leisure spaces, with the development comprising an area of over 70,000 m².

The 30,000 m² of external planting requires no conventional support from the local water authority, and relies completely on the recycled water production from the site. Energy usage levels are on a continual feedback loop to all apartments via 'smartboards' that register the usage and current demands.

Analysis of the planting, planter boxes, facade system and building orientation has shown a 35% reduction in the heat load on the building compared to conventional glass facade building forms. One Central Park achieves a Five-Star Green rating under the mixed use residential model, and a Six-Star Green rating for the precinct.

The sun reflection heliostat system is formally expressed and achieves an unexpected and memorable architectural impact. Its operation relies on a series of fixed and motorized mirrored panels on the roof of the 18-story West Tower, which track and reflect the sun to the cantilevered reflector frame at level 29 of the East Tower, which in turn directs light to areas otherwise in shadow. The system presents the first application of a solar technology, more typically deployed in remote Australia, within a residential and urban setting. The engagement with the path of the sun and the principles of permaculture ensures that this monumental tower is invested in both the operation and expression of sustainable living.

Credits and Data

Project title: One Central Park
Developer: Frasers Property Australia and Sekisui House Australia
Location: 2 Chippendale Way, Chippendale, New South Wales, Australia Completion: January 2014
Design Architect: Ateliers Jean Nouvel, Paris, France
Local Architect: PTW Architects, Sydney, Australia
Consultants: Watpac Construction (Builder), WSP (Basement Services), ARUP (Podium & Tower Services), Surface Design (Façade), Yann Kersale (Heliostat Lighting Designer), Turf Design And Jeppe Aagaard Andersen and Aspect Oculus (Landscape), Patrick Blanc (Green Wall Designer), Robert Bird Group (Structural Engineer), Davis Langdon (Quantity Surveyor), Hughes Trueman (Civil Engineers), City Plan Services (BCA Consultant), JBA Planners (Planning Consultant), Denny Linker & Co (Surveyor), Farra/ Karabiner Access (Façade Access Consultant), Acoustic Logic (Acoustic), Kennovations (Heliostat Mirror Consultant), Accessibility Solutions (Accessible Consultant), Masson Wilson Twiney (Traffic Consultant), GJames and JML (Façade Contractor), Junglefy (Green Wall Contractor), Palamont And Composite Industries (Planter Box Contractor), Tensile (Climbing Cable Contractor)
Gross Floor area: 67,626 m²
Project Area: 7,550 m²

UTS

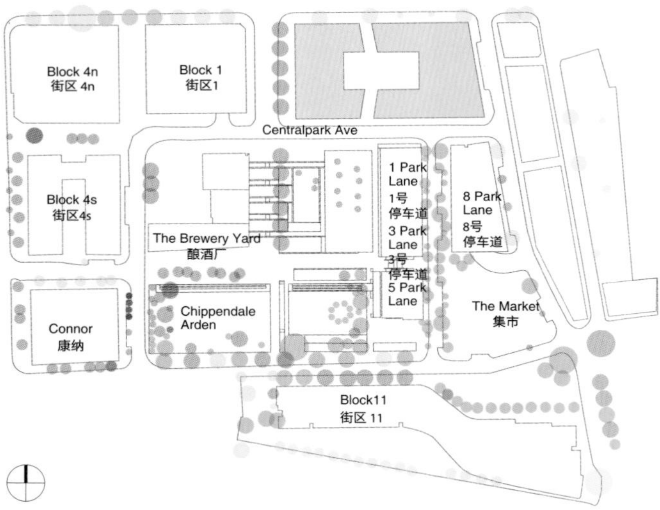

Site plan (scale: 1/4,000)／总平面图(比例: 1/4,000)

p. 191: A cantilevered heliostat redirects sunlight into the shaded areas. This page: Aerial view of One Central Park estate in downtown Sydney. Images on pp. 191-192 by courtesy of Frasers Property & Sekisui House.

第 191 页：日光反射悬臂将阳光引入阴影中的区域。本页：位于悉尼市中心的中央公园一号鸟瞰图。

中央公园一号是中央公园街区的地标性项目。中央公园一号的设计初衷和基本原则，是通过绿色开放空间和绿植的介入，实现可持续设计理念。其植物幕墙的设计贯穿整个高层住宅综合体，饱含它最初的设想并不断重申可持续设计的理念。

建筑由一栋 34 层高的东楼和一栋 18 层的西楼组成，两座塔楼共有 624 套大小、布局不同的公寓。塔楼下方是一个毗邻城市主干道的 5 层裙房，内部容纳了零售、商业和城市休闲设施，整个综合体占地面积超过 7 万平方米。

整个项目拥有 3 万平方米的室外绿植，并且这些绿植无需使用当地水务局的供水来灌溉，而是完全依靠场地产出的循环水。“智能板”会全程记录能源的使用情况和当前需求，并反馈给所有住户。

对种植、花箱、幕墙系统和建筑朝向的分析表明：相较于传统的玻璃幕墙建筑，该项目的热负荷降低了 35%。中央公园一号在住宅综合体类别下获得了“绿色之星”五星级评级，且整个规划区域获得了“绿色之星”六星级评级。

日光反射系统的使用达到了意想不到的效果，并对建筑产生了显著影响。日光反射系统的运作依赖于设在西塔楼 18 层屋顶的一系列固定电动镜面板，这些镜面板追踪阳光，并将光线反射到位于东塔楼 29 层的悬臂式反射框内，从而将光线引入原本处于阴影中的区域。这种太阳能技术通常在澳大利亚偏远地区使用，而本项目是该技术在城市住宅项目中的首次应用。对太阳轨迹的有效利用和对永续生活设计原则的遵循，确定了这座意义非凡的塔楼在可持续经营和表达方面的巨大投入与成果。

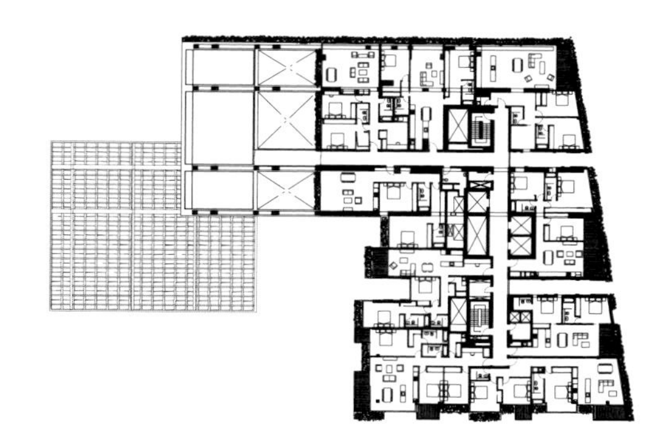

Heliostat upper floor plan／日光反射系统上层平面图

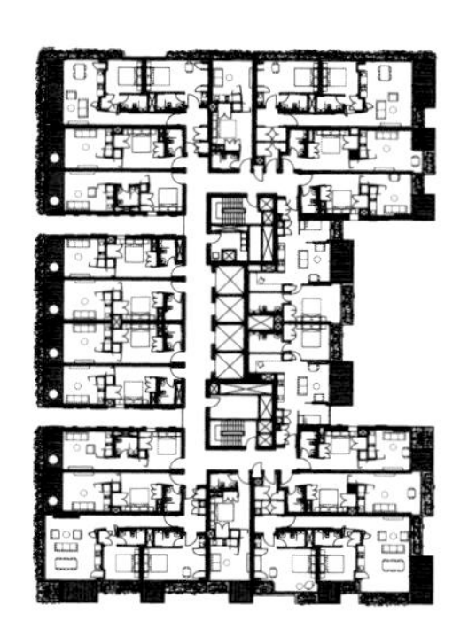

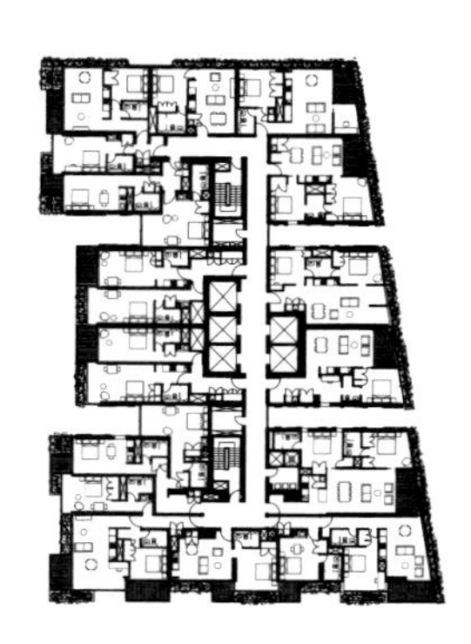

Typical floor plan (scale: 1/1,200)／标准层平面图（比例：1/1,200）

This page, above: View of the cantilevered rooftop terrace with expansive views of the city; below: View of the sunken courtyard and atrium that provides pedestrian connectivity within One Central Park. Opposite: Alternating vertical gardens and balconies help to bring a sense of nature into the residences. Images on pp. 194–197 by Simon Wood.

本页，上：悬挑结构的屋顶花园享有广阔的城市风光；本页，下：下沉广场和中庭确保了中央公园一号中人行路线的通达性。对页：交替的垂直花园和阳台有助于将自然引入住宅中。

This page: Vertical gardens comprising of a variety of indigenous and exotic climbing plants cover 50% of the building's façade.

本页：垂直花园由各种本土和外来引进的藤蔓植物组成，覆盖了建筑立面的 50%。

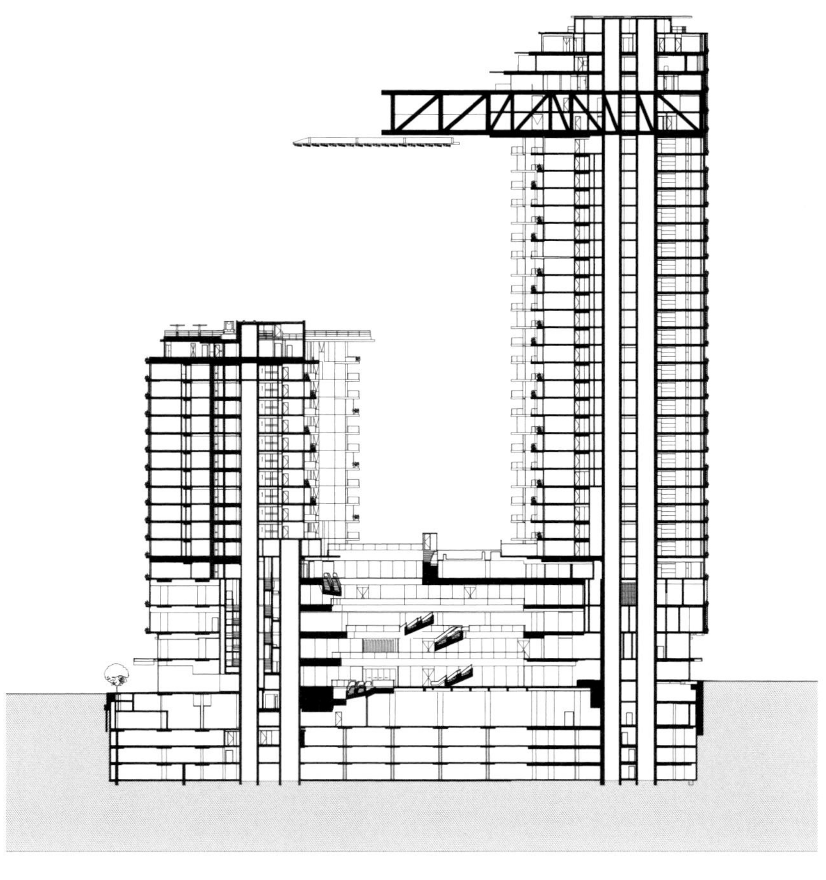

Section (scale: 1/1,200)／剖面图（比例: 1/1,200）

Water Urbanism in Australian Cities

Nigel Bertram, Catherine Murphy, David Mason
Melbourne / Brisbane / Perth

澳大利亚的水城市主义

尼格尔·伯特伦，凯瑟琳·墨菲，戴维·梅森
墨尔本／布里斯班／珀斯

Starting from the inclusive and interconnected understandings of urban ecology and the multiple time scales of geomorphology and environmental history, these projects seek to understand more about the conditions of our urbanized places in order to conceive of how they might evolve into the future – in ways that are more water sensitive, more responsive to changing natural and human-made phenomena, more enjoyable for more people, and more resilient in their ability to absorb the shock of new and unforeseen events. To develop a process for this research we started by turning our attention towards the past. From this seemingly counterintuitive viewpoint we placed ourselves as designers in a position to discern particular types of knowledge and to listen to particular other voices, that might have been overlooked, dismissed or simply forgotten in the ongoing process of continual technological improvement, development and competitive change that characterises modern urban space and its planning. Such voices include those of local cultural memory, indigenous knowledge and practice, traditional techniques from other cultures, complex animal and plant systems, and knowledge of the subsurface realm including geology, groundwater and contamination. Focusing our study of urban environments through the lens of water provides us an interface with all of these. Water is dynamic and enduring – an agent which can cross time and space, transcending physical and cultural boundaries.

These projects were part of the Cooperative Research Centre for Water Sensitive Cities (CRCWSC), an Australian research centre that was established in 2012 to help change the way we design, build and manage our cities and towns by valuing the contribution water makes to economic development and growth, our quality of life and the ecosystems of which cities are a part. Through research projects, the CRCWSC brings together many disciplines, world-renowned subject matter experts, and industry thought leaders who want to revolutionize urban water management in Australia and overseas. As part of the CRCWSC, this project explored how a design-led approach could synthesize a range of complex issues towards delivering water sensitive urban design for precinct scale, redevelopment sites in Melbourne, Perth and Brisbane. The project developed a range of processes towards localized, small-scale water systems that could contribute to climate change resilience and intensification issues in cities.

Melbourne – Arden Macaulay Island City

This design vision is for the regeneration of the low-lying, post-industrial Arden Macaulay precinct in central Melbourne. Referencing the hybridity of other altered lowlands, this vision imagines a future where dynamic natural and human-made environments co-exist in a dense urban setting. The area is prone to intensified flooding and is threatened by sea level rise, as the deep structure of the underlying city reasserts itself through pressures of climate dynamics, as well as through the urbanization of the upper Moonee Ponds Creek Catchment.

Arden Macaulay sits in the larger environmental context of the Victorian Southern Lowlands, a physiographic region south of the Great Dividing Range, which includes significant previously swampy sites, in both urbanized and rural areas. Sites of regional landscapes were documented to

这一系列项目结合地质年代和环境史，从全面调查都市生态开始，试图进一步了解城市化的情形，探索在未来如何提高城市水敏性，以更好地应对不断变化的自然环境和人为现象，让更多人感到更愉悦，同时更有力地消化不可预见的新事件的冲击。研究之初，我们就抛开了作为设计者的直觉，将关注点放在已经发生、存在的事物上，分辨特定的知识类型，倾听特定的另类声音，例如当地的文化记忆、原住民的知识和实践、源于其他文化的传统技能、复杂的动植物系统、有关地表下方（包括地质、地下水、污染）的信息。这些声音，很可能随着当代城市空间的营造及规划，在持续的技术进步和竞争变化中被忽视或遗忘。透过水研究城市环境，我们会看到以上所有问题的相通之处，因为水是流动且持久的——它能够跨越时空，超越物质与文化的边界，成为万千事物的媒介。

这些项目所属的“水敏性城市合作研究中心”（CRCWSC）成立于2012年，是澳大利亚的一个研究中心，旨在评估水对经济增长、生活质量、城市所处生态系统的贡献，进而帮助我们改进城镇规划、建造和管理的方式。通过研究项目，CRCWSC联结起了世界知名的跨学科专家，以及希望在澳大利亚乃至全世界改革城市水管理的行业思想领袖。作为CRCWSC的一部分，这些项目揭示了在墨尔本、珀斯、布里斯班的再开发地区，以设计为主导的方法如何应对水敏性城市设计中的各种复杂问题。项目还摸索出建立本地化小规模水系统的流程，能为应对城市气候的剧烈变化、推动其恢复做出贡献。

墨尔本——阿登麦考利岛屿城市

这是为了重整墨尔本市中心低洼地兼老工业区阿登麦考利而做的设计构想。考虑到其他低洼地的混杂性，该方案设想了一个动态自然和人造环境在密集城市中共存的未来。这一地区本就易发洪水，且受到海平面上升的威胁；随着近年气候的剧烈变化、曼莉河上游的城市化，城市地表下方深层结构的重要性更加凸显。

阿登麦考利位于大分水岭以南、维多利亚州的南部低地，拥有许多重要的沼泽地，它们分布在城市与乡郊。有记录的沼泽地包括康达湖、温顿湿地、麦克里德沼泽、鹅颈沼泽等，它们都经过了细微的基础改造。阿登麦考利附近还有曾经十分广袤的西墨尔本沼泽，如今已扩张为墨尔本低地，占据了一度草木丛生的间歇性河口湿地。研究这些环境都有助于了解场地的基本性质。

这一项目正是受到环境的启发，进而给重整池塘、沼泽连片的阿登麦考利提供了策略。从漫长的原住民历史来看，这里曾是进行聚会、交流的地方，生态富饶；而在相对短期的西方式工业化进程中，受污染的沼泽和断流的水道因经济利益被填埋或疏浚。项目恢复了阿登麦考利一些固有的生态和文化价值，同时保留了它的近代工业遗存。当河水重新流经生机勃勃的岛屿湿地并服务更多人的时候，自然环境和建成环境已然准备好了迎接未来的变化。

利用这种方法，可以发展出新的城市结构和建筑类型，凸显城市的韧性和适应性。我们希望，在了解整体状况（洪水强度和频率增加、旱涝循环、海平面上升、湿地作为缓冲带）的基础上，创造与时俱进的都市，并通过新的生态系统，给开发较少的偏远地区提供支持。

布里斯班——诺曼河防洪景观，抬升地面

在昆士兰的布里斯班内城、诺曼河集水区，高度城市化的郊区与中央商务区仅隔一条布里斯班河，被认为很适合容纳都市扩张。但由于紧邻一条汇入河道的小支流，那里极易受

inform the Arden-Macaulay design: Lake Condah, Winton Wetlands, Macleod Morass and Gooseneck Swamp, all of which are small, acupuncture-like infrastructural modifications. Closer to Arden Macaulay, the previously vast West Melbourne Swamp and its extension into the Melbourne Lowlands, a region of once marshy, estuarine intermittently wet space, were studied to understand the site's underlying nature.

These parallel environments informed this architectural project, which proposes strategies to regenerate Arden Macaulay, a place that was once a swamp and chain of ponds. Its longerterm Indigenous history was a place of meeting, of exchange, of environmental abundance. Its shorter-term history is one of western industrialization, where polluted swamps and broken waterways were eradicated or channelized for economic gain. This proposal restores some of the inherent ecological and cultural values of Arden Macaulay, while also celebrating its more recent industrial heritage. By re-allowing water to flow through a flourishing wetland habitat of islands for growing numbers of people, the natural and built environments are prepared for future change.

Through this approach, new urban frameworks and building typologies can evolve that address urban resilience and adaptation: increased flood intensity and frequency, droughtflood cycles, sea level rise (wetland as buffer / shock absorber) and a city evolving over time to work more consciously in tandem with its underlying natural structure while also, through a novel ecosystem, support those less modified further afield.

Brisbane – Norman Creek Inhabited Floodscape, Raised Ground

The highly-urbanized suburbs within the Norman Creek catchment of inner Brisbane, Queensland, are separated from the CBD only by the Brisbane River and have been identified as suitable for accommodating substantial urban growth. The catchment forms around a small tributary flowing to the river and is susceptible to major flooding such as the 2011 event that devastated sections of the city. Since the previous major flood event of 1974, substantial urban infill development has occurred within the floodplain of the Brisbane River.

This design project, located at Turbo Drive Isle in the lower part of the catchment, is a response to this flooding and looks towards a sustainable urban form for the city. A new, raised ground provides safe connectivity within the urban cluster and with the rest of the city. Habitable and service spaces, public and private, open and enclosed can develop both above and below the safe ground, which is set above the 2011 flood level – to cater for a 30% increase in water level due to climate change.

The three-dimensional platform, between the raised and floodable grounds, hosts water-proof commercial and communal spaces or storage rooms and carparking that are "sacrificial", providing room for flood waters. The majority of vehicular movement is retained below the platform level in order to clear the way above for pedestrian movement. A network of routes links public transport to mixed use nodes, down to the natural green corridor along the creek, conveying water flows to retention and filtering ponds and "urban sponges".

These strategies improve the water quality by trapping sediments, filtering out pollutants and absorbing nutrients that would otherwise flow into the creek. Flanking it, the wetland serves as a buffer zone, a liminal boundary that expands the creek bed during flooding. The concept of water urbanism deployed in the project, transcends mere landscaping treatment, establishing instead a dialectic between natural and constructed environments, which are shaped by their relation with water. The power of water in the making of contemporary cities must be reinstated through design strategies, at both urban and building scale, that conceive living with water as a privileged condition rather than a threat.

Perth – The deep waters of Perth

Perth is a young city on Australia's ancient edge. The oldest rocks in the world are found in this region, further inland on the Yilgarn Block. Below the Yilgarn escarpment is Perth, a city of two million spreading along the Swan Coastal Plain. Extending beneath the surface of this city are large reserves of water stored in deep aquifer layers. When the groundwater emerges on the surface, a tapestry of rivers, creeks and seasonal wetlands form some of

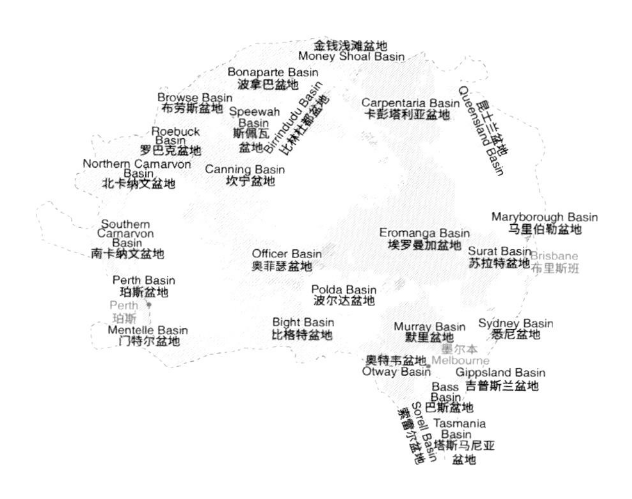

Blocks and basins of the Australian continent (key sedimentary basins highlighted in pink).
澳洲大陆的地块与盆地（主要沉积盆地以粉色标示）

到类似2011年洪涝等灾害的破坏。事实上，自1974年特大洪灾以来，布里斯班河的洪泛区就出现了大量填充式开发。

这一位于集水区下游涡轮岛的项目，正是对2011年洪涝的回应，也是对城市可持续发展形态的展望。抬升的新地面建在2011年洪水水位上方，以应对气候变化带来的30%的水位上升，同时为城市中心与其余地区提供安全的连接。地上或地下都可以进行公共或私人的开发，形成开放或封闭的空间，用于居住或商业服务。

抬升地面和可淹地面之间还有一个立体平台，内部可用作防水的商业空间、公共空间、储藏室、停车场，必要时可以“牺牲”于洪水中，用于滞洪。大部分车行道仍在平台下方，为上方的人行道留出场地。整个线路网连接起公共交通和混合用途节点，并且一直通到诺曼河畔的自然绿色走廊，水也经由这条小河流向蓄水池、过滤池和“城市海绵”。

如此一来，通过截留沉积物、过滤污染物、吸收原本会流入诺曼河的营养物，水质将会得到改善。另外，位于河流两侧的湿地在洪水泛滥时可以延伸河床，起到滞洪的作用。该项目中的水城市主义概念超越了单纯的景观规划，建立了自然环境和建成环境之间的对话，也受到环境与水的关系的影响。建设当代城市时，必须从城市和建筑两方面重新认识水的力量，将与水共生视为一种特性而非风险。

珀斯——珀斯的深水

珀斯是一座年轻的城市，坐落于澳大利亚历史悠久的边缘地带，世界上最古老的岩石就发现于该地区的内陆——伊尔岗地块。伊尔岗的断崖之下，正是珀斯这座拥有两百万人口、沿着天鹅海岸平原展开的城市。城市地下深处的含水层储存有大量水资源，当地下水出现在地表时，河流、小溪和季节性湿地便构成了世界上最具地方特色、生态最丰富的景观之一。这些水文系统对原住民而言非常重要，蕴含于当地努恩嘎文化传统之中。

然而，城市化给这些水系统的健康带来了巨大的挑战，不适当的郊区开发损害了珀斯一直以来的地下水、水文和生态环境。不过，良好的住房设计可以有助于它们的恢复和维护。西澳大学建筑系的学生们就透过城市的深层和表层，研究了一些对地下水敏感的设计对策。

不同于珀斯郊区开发中普遍存在的肆意挖填，这些设计对水的状况更加敏感。它们引入轻型工法，在湿地上建造和居住，保护已有的植被和水文。中等密度的住宅规划还为公共空间提供了场地，居民的景观维护意识得到增强。不论哪个案例，皆改善了生态性和水敏性，适当提高了郊区的人居密度。此外，在探寻与水共处（而非消除或隐藏水）的过程中，对成本高昂的淡化水的依赖（目前珀斯一半以上的用水需求都源于淡化水）也有所减少。项目中，场所本身被设计为集水区，能够收集水、改变其流向，有利于形成更宜居、更健康、可改善微气候的城市景观。相信在未来，这种对地下水敏感的设计将广泛应用于世界众多的地下水城市。

the most endemic and ecologically rich landscapes in the world. These hydrological systems are of great importance to Indigenous people, embedded in the traditions of the local Nyungar culture.

Urbanization presents a great challenge to the health of these water systems. Inappropriate suburban development has exploited Perth's groundwater, the hydrologies and ecologies they sustain. However, good housing design can become a catalyst for recovery, protection and conservation of these environments. Groundwater sensitive design responses have been explored by architecture students at the University of Western Australia, through the lens of the city having depth as well as surface.

Rather than the careless cutting and filling of land which has become commonplace across Perth's sprawling suburban development, these design approaches are more sensitive to water conditions. They involve lightweight methodologies of building and dwelling above wetlands, protecting existing vegetation and hydrology. These medium-density housing schemes create places with shared communal connections encouraging custodianship of landscape. In every case, they improve ecological and water sensitivity while also increasing density within the suburban context. By exploring ways of dwelling with water rather than erasing or hiding water, these schemes reduce dependencies on costly desalinated water (now serving more than half of Perth's water needs). They treat their sites as catchments, and are able to collect and reroute water to create urban landscapes that enhance livability, health and microclimate. Such groundwater-sensitive design will be across many of the world's groundwater cities in the future.

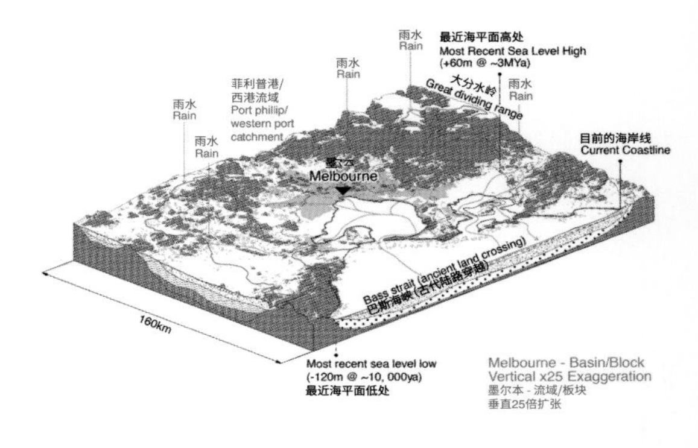

Axonometric block / basin diagram of Brisbane's bowl structure
轴测图/墨尔本沼泽洼地示意图

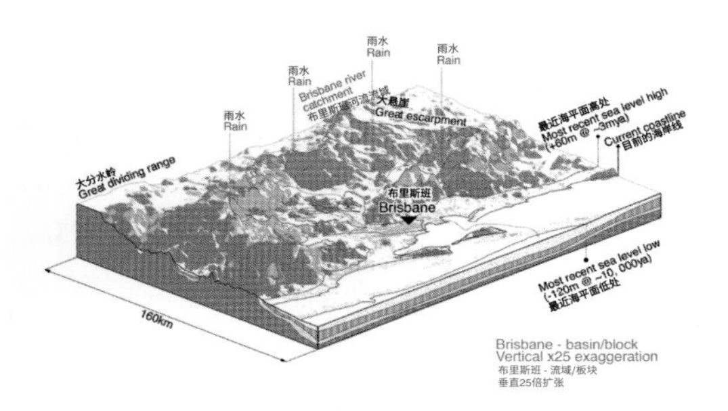

Axonometric block / basin diagram of Melbourne's sunkland structure
轴测图/布里斯班下沉式洼地示意图

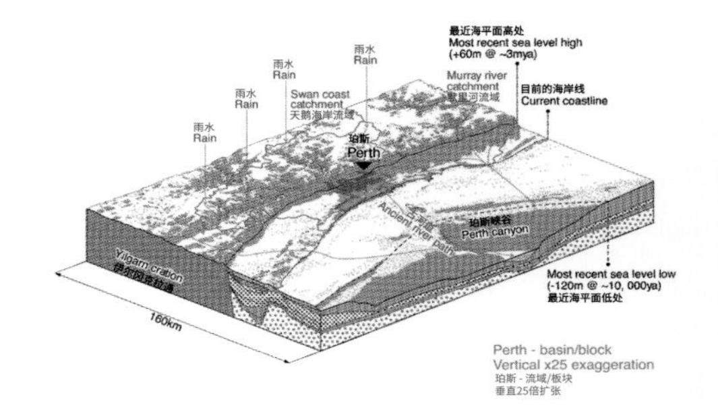

Axonometric block / basin diagram of Perth's coastal plain structure
轴测图/珀斯海岸平原洼地示意图

Melbourne – Arden Macaulay Island City

墨尔本 - 阿登麦考利岛屿城市

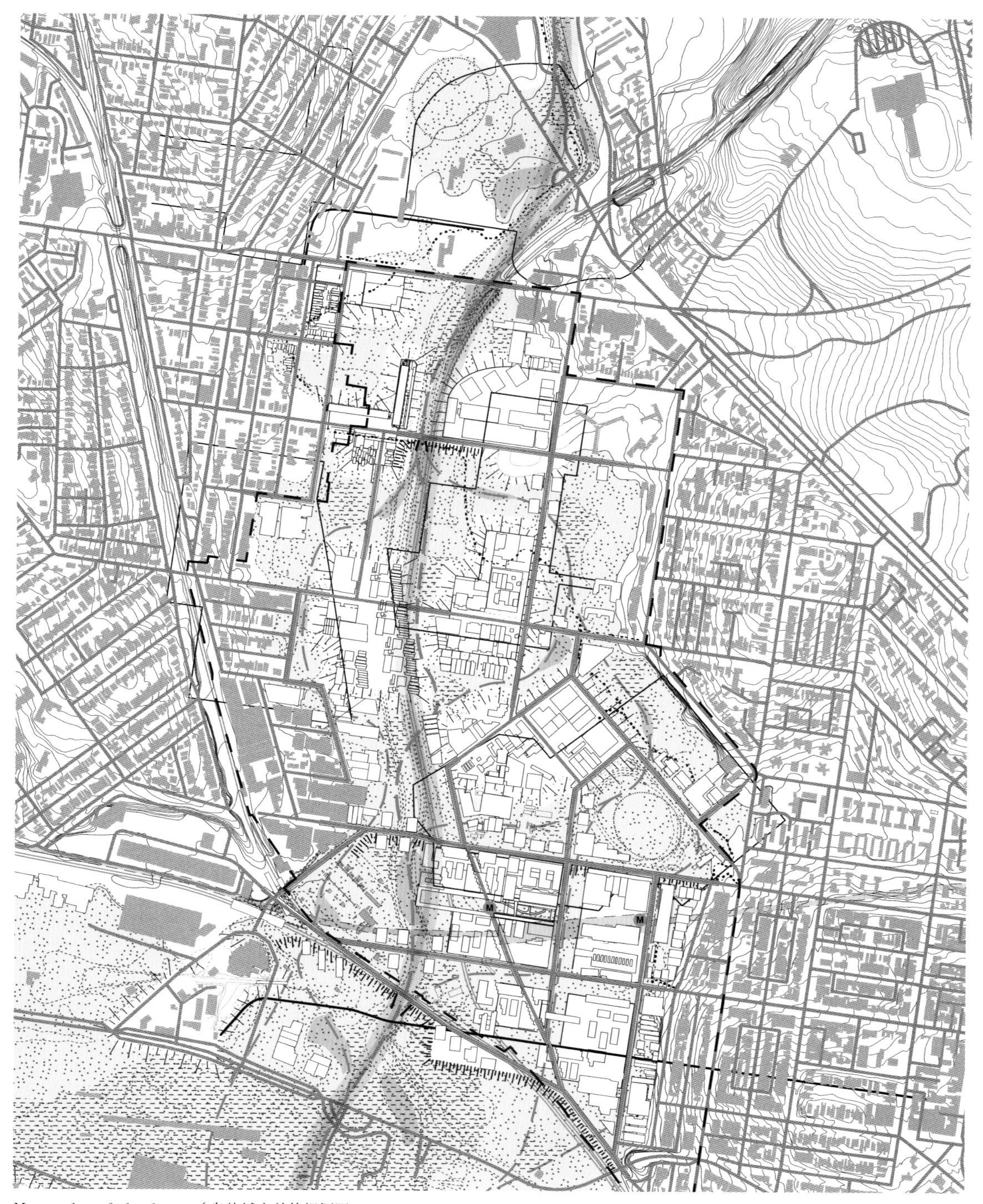

Master plan of island city ／ 岛屿城市总体规划图

Axonometric drawing of Island City (below: during flood)
岛屿城市的轴测图（下：洪水期间）

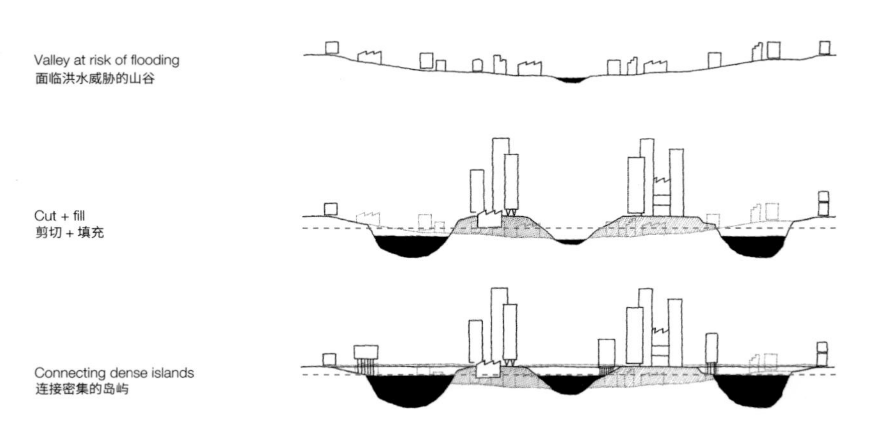

Section diagram of Island City
岛屿城市剖面图

Braided water catchment
分叉型集水区

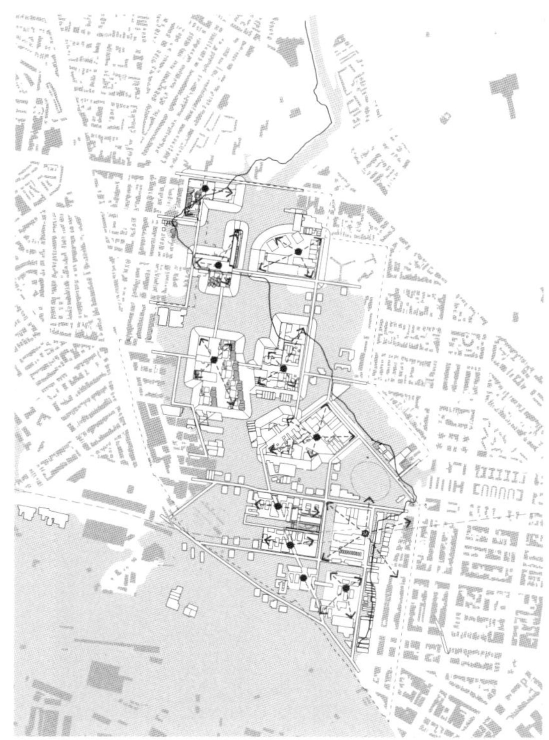

5 minute island cities
五分钟步行圈内的岛屿城市

This page, above left: View from chain of ponds / new public realm to high- density island. This page, above right: Macaulay Common: A groundwater-fed swamp covered with couch grassland and absorbent peat moss acts as a retarding basin with perpendicular connections across the creek.

本页，左：从连片的池塘望去 / 高密度岛屿带来新的公共场所；本页，右：麦考利公园：覆盖着卧草与吸水泥炭藓的沼泽地，由地下水补给，与河道垂直相连，起到滞洪的作用。

Brisbane – Norman Creek Inhabited Floodscape, Raised Ground

布里斯班 – 诺曼克里克人居住的洪水景观，高架地面

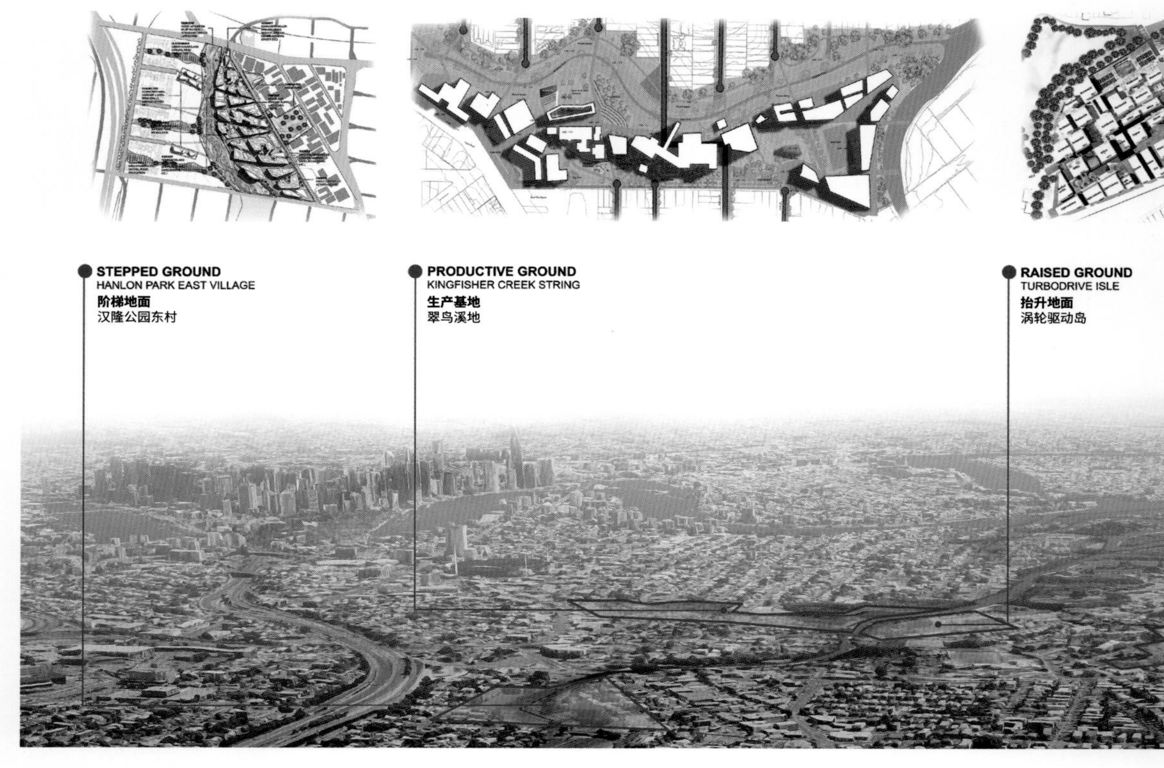

Norman Creek Catchment masterplan locating Turbo Drive Isle project.
涡轮岛项目所在的诺曼河集水区总规划图

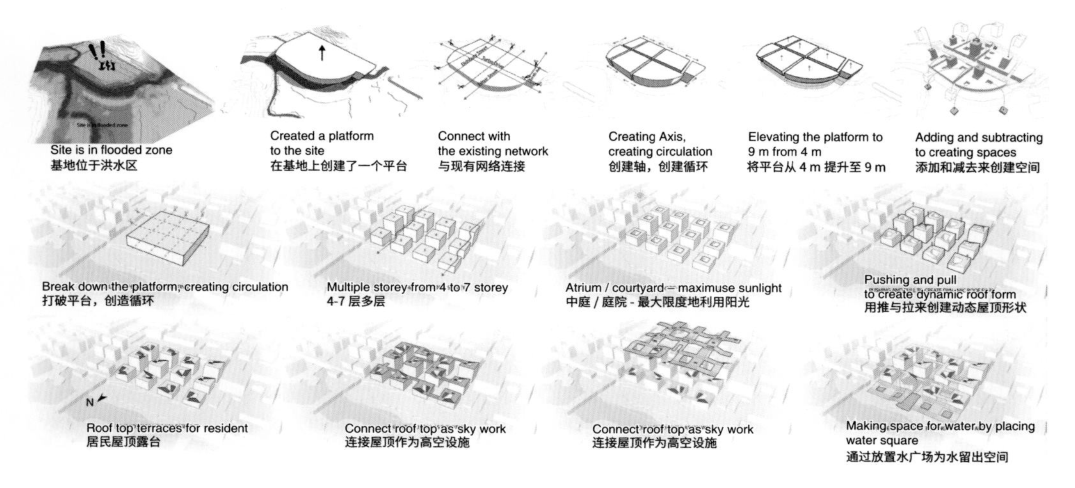

Raised Ground, Drive Isle, concept diagrams. Project by Charisa Chan Yan Yan, School of Architecture, The University of Queensland.
抬升地面在涡轮岛的概念图

COMMON GROUND
COORPAROO TRIANGLE
公用地面
库帕罗三角

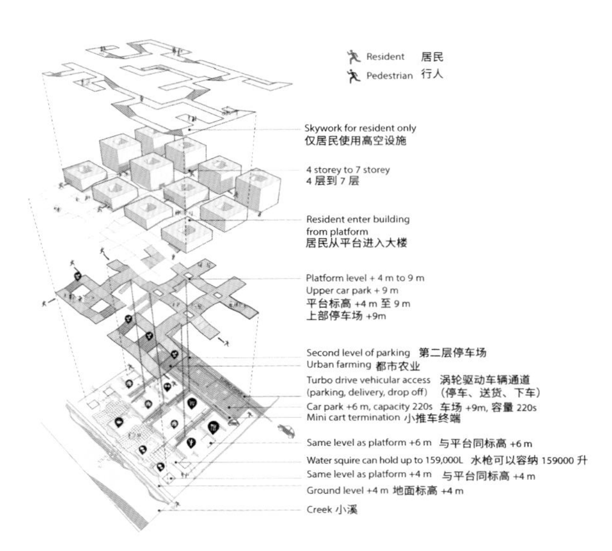

Raised Ground, layered systems /抬升地面，分层系统

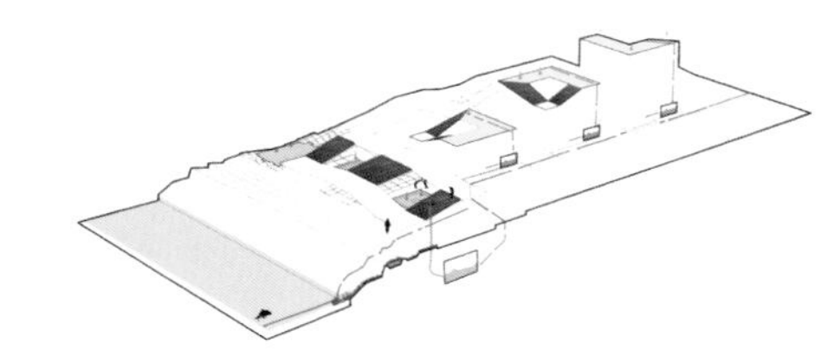

Raised Ground, urban sponge strategy /抬升地面，都市的海绵策略

Raised Ground, aerial perspective /抬升地面，俯瞰图

Raised Ground, Creek interface /抬升地面，与河道的交接面

Perth – The deep waters of Perth

珀斯 – 珀斯的深水

This page, above right: Perth's urban area stretches along the Swan Coastal Plain between the Indian Ocean and the Yilgarn Block, forming a clear topography.

本页，右上：珀斯的都会区沿着印度洋和伊尔岗地块之间的天鹅海岸平原延展。

This page, below left: Rich chains of wetlands once extended across the city, surface expressions of the groundwater below.

本页，左下：地下水涌出地表，形成在全城蔓延的大量湿地带。

This page, below right: The aquifers form mountains of water beneath the city, posing significant challenges and opportunities for Perth's future urbanization.

本页，右下：含水层在城市地下形成“水山”，给珀斯未来的城市化带来重大的挑战和机遇。

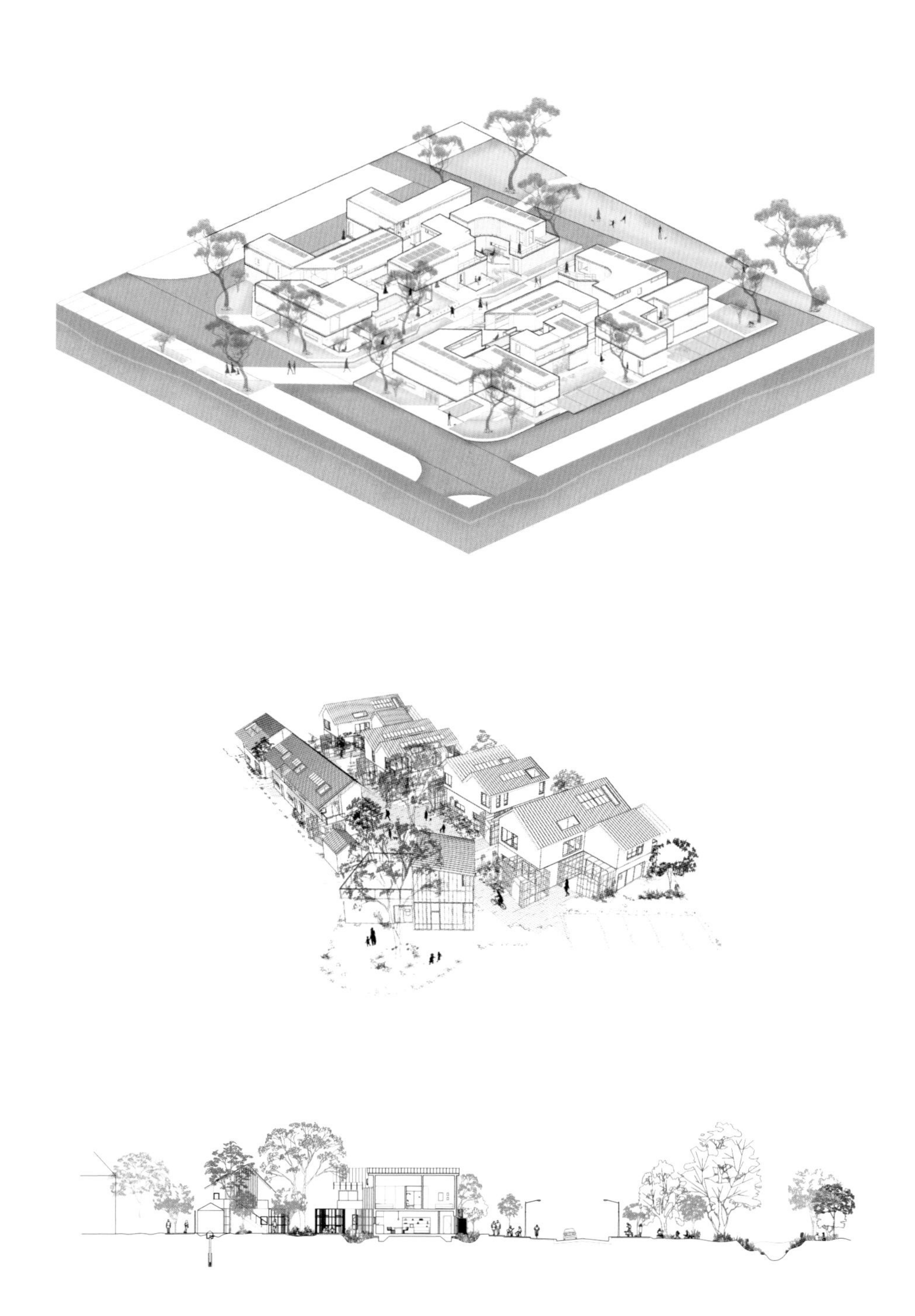

Groundwater sensitive design responds to water as a vital element of urban life.
对地下水敏感的设计中，水被视为都市生活的一大要素。

Architects Profile

建筑师简介

John Wardle established his architectural practice in Melbourne and has led the growth of the practice from working on small domestic dwellings to university buildings, museums and large commercial offices. John has an international reputation as a design architect and has developed a design process that builds upon ideas that evolve from a site's topography, landscape, history and context and the client's particular aspirations and values.

约翰·沃德尔在墨尔本创立了他的建筑事务所，从承接本地的小型住宅设计逐步发展到大学教学楼、博物馆、大型商业办公楼设计，并带动了建筑实践的发展。约翰是一位享有国际声誉的建筑师，他开发了一种设计流程，其理念基于场地地形、景观、历史和背景以及客户的特定愿望和价值观。

Baracco+Wright is led by its two founding directors, **Mauro Baracco** and **Louise Wright**. Their architectural practice combines the academic and practice world and is shifting more and more towards landscape based approaches. They are the Creative Directors in collaboration with artist Linda Tegg of the Australian Pavilion at the 16th International Venice Biennale di Architettura 2018 with their theme “Repair”.

巴拉科与赖特建筑师事务所由其两位创始合伙人**莫罗·巴拉科**和**露易丝·赖特**领导。他们的建筑实践结合了学术和实践领域，并越来越转向基于景观的设计方法。在2018年第16届国际威尼斯建筑双年展上，他们作为创意总监，与艺术家琳达·泰格合作完成了主题为“修复”的澳大利亚馆。

Michael Trudgeon has practiced across the disciplines of architecture, industrial design and communication design since 1983. He is the Director and senior designer at Crowd Productions. Crowd Productions is a design studio focused on creating and curating customer and staff experiences in physical space. They have 20 years of experience designing innovative retail and entertainment spaces for banks, retailers, health centres, cinemas and museums.

迈克尔·特劳根自1983年以来一直从事建筑设计、工业设计和视觉传达领域的跨界实践工作。他是群作工作室的负责人兼高级设计师。群作工作室是一个设计工作室，专注于在物理空间中创造和策划客户及员工的体验。他们为银行、零售商、健康中心、电影院和博物馆设计创新的零售和娱乐空间，拥有20年的经验积累。

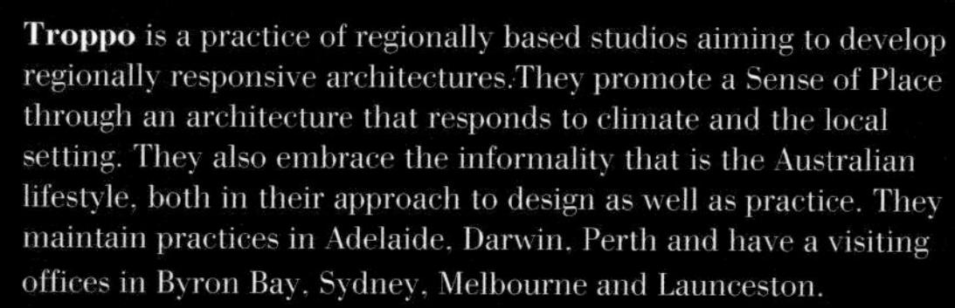

Troppo is a practice of regionally based studios aiming to develop regionally responsive architectures.They promote a Sense of Place through an architecture that responds to climate and the local setting. They also embrace the informality that is the Australian lifestyle, both in their approach to design as well as practice. They maintain practices in Adelaide, Darwin, Perth and have a visiting offices in Byron Bay, Sydney, Melbourne and Launceston.

特罗普建筑师事务所是一家地域性的工作室，旨在建造能反映地域性的建筑。他们通过设计响应气候和当地环境的建筑来提升“场所感”。他们在设计方法和实践中都采纳了澳大利亚人生活方式中的随意性。他们在阿德莱德、达尔文、珀斯持续开展业务，并在拜伦湾、悉尼、墨尔本和朗塞斯顿设有访问接待处。

Brit Andresen studied architecture in Norway (NTH). Her private practice in Cambridge U.K. and partnership with Peter O'Gorman in Brisbane have resulted in design research and built works that have been published and exhibited internationally, including Berlin 2007, Venice Biennale 2001 and 2010. She has taught architecture at the University of Cambridge, Architectural Association London, School of Architecture and Urban Planning UCLA.

布丽特·安德烈森在挪威皇家理工学院学习建筑。她在英国剑桥的个人实践，以及在布里斯班与彼得·奥戈曼的合作，使她的设计研究和建筑作品多次被收录在国际出版物中并参与国际性展览，其中包括2007年的柏林展，和2001年、2010年的威尼斯双年展。她曾在剑桥大学、伦敦建筑联盟、加州大学洛杉矶分校建筑与城市规划学院教授建筑学。

Richard Leplastrier is an architect, teacher and lives modestly with his family in a beautifully crafted home he designed overlooking Pittwater, North of Sydney, New South Wales, Australia. Richard grew up in Perth, Hobart and Sydney and studied architecture at the University of Sydney. He worked for Jørn Utzon from 1964 to1966 and spent 18 months in Kyoto studying traditional Japanese architecture with Professor Masuda Tomoya and later worked with Kenzo Tange.

理查德·莱普拉斯特里尔是一位建筑师、教师。他和家人住在由他本人设计的一座建造精致的住宅中，过着舒适的生活。他的住宅俯瞰着澳大利亚新南威尔士州悉尼北部的皮特沃特河口。理查德在珀斯、霍巴特和悉尼三地长大，并在悉尼大学学习建筑。1964年至1966年，他为约恩·乌松工作，后在京都师从增田友也教授进行了18个月的日本传统建筑学习，曾与丹下健三共事。

Wendy Lewin was born in Sydney in 1953. She obtained her architectural degree from the University of Sydney. She established her private firm in 1993, working as a sole practitioner and in association with Glenn Murcutt. Glenn Murcutt was born in London in 1936. He received his Diploma of Architecture from University of New South Wales Technical College in Sydney. He received the Pritzker Architecture Prize in 2002.

温蒂·卢因于1953年在悉尼出生。她在悉尼大学获得建筑学学位。1993年，她成立了自己的私人公司，以个人执业者的身份工作，并同时与格伦·马库特合作。**格伦·莫卡特**于1936年出生于伦敦。他在悉尼的新南威尔士大学工学院获得了建筑文凭，曾于2002年获得普利兹克建筑奖。

Andrew Burns graduated from the University of Sydney in 2004. Since its establishment in late 2007, his practice has undertaken residential, cultural, community and public projects. The practice's approach seeks to combine social engagement with design excellence, and is characterised by precise geometry and material exploration.

安德鲁·伯恩斯于2004年毕业于悉尼大学。他的事务所自2007年末成立以来，接手的项目涉及住宅类、文化类、社区类及公共建筑类。该事务所旨在将社交互动与卓越设计相结合，并以精确的几何形状和材料探索为特征。

Bud Brannigan is the director of Bud Brannigan Architects, based in Brisbane, Australia. The small practice, begun in 1993, has been honoured with several awards from the Australian Institute of Architects. In 2007, Bud Brannigan also received a Churchill Fellowship to study regional museums in the USA, and in 2011, the practice was shortlisted to design the new Australian pavilion for the Venice Biennale.

巴德·布兰尼根是巴德·布兰尼根建筑师事务所的负责人，该事务所总部位于澳大利亚布里斯班。这家小型事务所始于1993年，并获得了澳大利亚建筑师学会的多项殊荣。 2007年，巴德·布兰尼根还获得了丘吉尔奖学金，用以研究美国的地区博物馆。2011年，该事务所入围了威尼斯双年展新澳大利亚馆的设计。

Based in Fremantle, Western Australia, Officer Woods (OW) Architects is led by its two founding directors, **Jennie Officer and Trent Woods**. OW has a particular interest in working with diverse organisations and users. OW is committed to cultural engagement and innovation in all projects, using prosaic and well understood means to make something remarkable. Recently they won the national Sir Zelman Cowen Award for Public Architecture in 2017 for the East Pilbara Arts Centre.

总部位于西澳大利亚州弗里曼特尔的奥菲瑟·伍兹建筑师事务所（以下简称OW）由其两位创始合伙人**珍妮·奥菲瑟**和**特伦特·伍兹**领导。OW对跟不同的组织和用户合作十分感兴趣。OW致力于在所有项目中实现文化参与和创新，并使用平常且广为人知的方法来完成令人瞩目的设计。近期，他们凭借东皮尔巴拉艺术中心赢得了2017年"国家塞尔曼·考恩爵士公共建筑奖"。

John Choi is partner of CHROFI. established in 2000. the practice's founding design. TKTS. has been widely recognised for its design excellence and innovation. from fields as varied as planning. architecture. branding. public space and tourism. He is Adjunct Professor of Architecture at University of Sydney.

约翰·崔是2000年成立的克洛菲的合伙人，该事务所的创始设计作品纽约时代广场售票厅（TKTS），因其杰出的设计和创新性受到了来自包括规划、建筑、品牌、公共空间以及旅游等各领域的广泛认可。约翰·崔也是悉尼大学建筑系的兼职教授。

Ray Brown has been a Director of Architectus for over 20 years and plays an active role in developing the strategic direction of the practice. Leading multiple project teams. He drives the Architectus teams to effectively express the materiality and tectonics of how spaces are put together – without hinging on particular aesthetic preconceptions. This relative freedom of response enables truly unique structures and spaces to evolve.

雷·布朗担任Architectus建筑师事务所的负责人已超过20年，对该事务所战略方向的发展发挥着积极的作用。他领导多个项目团队，并推动团队有效表达空间构成中的物质性和构造性，而不依赖于特定的审美观念。这种相对自由的应对方法使真正独特的结构和空间得以逐步形成。

Tzannes is an Australian studio for architecture. urbanand integrated design based in Sydney. lead by five directors. **Alec Tzannes. Jonathan Evans. Ben Green. Chi Melhem and Mladen Prnjatovic**. The practice's creative thinking and innovative. sustainable and enduring architecture has established their reputation as a leader in the field.

哲纳司建筑师事务所（Tzannes）是一家总部设在悉尼的澳大利亚建筑设计及城市整体规划事务所，由五位负责建筑师领导：**亚力克·哲纳司，乔纳森·埃文斯，贝恩·格林，池伊·梅尔赫姆**和**穆拉登·普托维奇**。该事务所的创造性思维，以及他们的建筑中表现出来的创新性、可持续性和耐久性，确立了他们在该领域的领先地位。

Monash Urban Laboratory is a research unit within the Faculty of Art. Design & Architecture (MADA) at Monash University. Melbourne. It has the ambition of making a significant contribution to some of the most pressing urban issues facing cities and regions. This wider project is led by **Nigel Bertram** and **Catherine Murphy** at Monash. in collaboration with University of Western Australia and University of Queensland.

莫纳什城市实验室是隶属于墨尔本莫纳什大学艺术设计与建筑学院（MADA）的一个研究单位。它追求的目标是为城市和地方面临的一些最为紧迫的城市问题做出重大贡献。这个涵盖面极广的项目由莫纳什大学的**尼格尔·伯特伦**和**凯瑟琳·墨菲**领导，并与西澳大利亚大学和昆士兰大学合作进行。

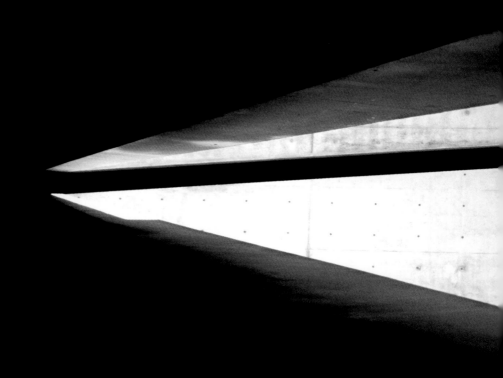

中日邦交正常化50周年纪念项目　日本国际交流基金会赞助项目

致敬20~21世纪传奇建筑家

全解建筑世界里的光影挑战

[illegible]/结构 六卷/印刷 四色全彩/装帧 精装/发行范围 中国/内容监修 安藤忠雄建筑研究所/特别支持 日本新建筑社/书籍策划 文筑国际 IAM国际建筑联

[illegible] 铃木博之 彼得·艾森曼 松叶一清 中川武 三宅理一 朱涛 马卫东等/本书相关消息敬请关注官方微信公众号"安藤忠雄之家"

Spotlight:

Shanghai Poly Grand Theater

Tadao Ando Architect & Associates

特别收录：

上海保利大剧院

安藤忠雄建筑研究所

The site is located in Shanghai suburb, Jiading, where a rapid urban development takes place now. This is a project of a cultural complex including an opera house, which is currently under construction to become a cultural center of a 100km² new town .

Shanghai, which was called a "magical city of the East", has become an international city so rapidly since the opening of a port in the middle of nineteenth century and has been a culture-importing base that transmits modern town civilization to all over China. The source of energy exists in the crash of unlike things such as the West and the East and tradition and modernism. Since cities always struggle to be stimulating and intense, required architectural forms are symbolic, or almost kitschy ,and figurative. In such Shanghai I am attempting to create an architecture that contains the city's passion inside it and not on the external form of the building.

Poly Theater + Commercial & Cultural Center is situated at the center of a green-filled park that possesses a large artificial lake on the east side of a new town. An opera house facing the artificial lake poses a plain rectangular volume in the size of 100m x 100m in plan and 30m in height. In the volume cylindrical void spaces in 18m diameters are inserted from various angles from top to bottom and left and right. The three-dimensional spaces developing univer-sally sustain the building's composition and a 1,600 seats-capacity main hall is planned at the center.

The mixture of solid and void and tubes and cubes create a range of characters of places, so that the architecture embraces complicated and diverse spaces in a simple framework.

Text by Tadao Ando Architect & Associates

Credits and Data
Location: Shanghai, China
Design: 2009.3-2010.3
Construction: 2010.8-2014.7
Structure: SRC, steel
Function: Theater
Site area: 30,235.00 m²
Building area: 12,450.00 m²
Total floor area: 54,934.00 m²

pp. 216–217: Side facade of Shanghai Poly Grand Theatre. Image by Shigeo Ogawa. This page: Discussion manuscript. Opposite: Shanghai Poly Grand Theatre lights up.p. 220: Fair-faced concrete walls in the cylinder space define the corridors and steps.

第 216-217 页：上海保利大剧院侧立面。本页：研讨手稿。对页：华灯初上的上海保利大剧院。第 220 页：圆筒空间中的清水混凝土墙定义出廊道以及阶梯。

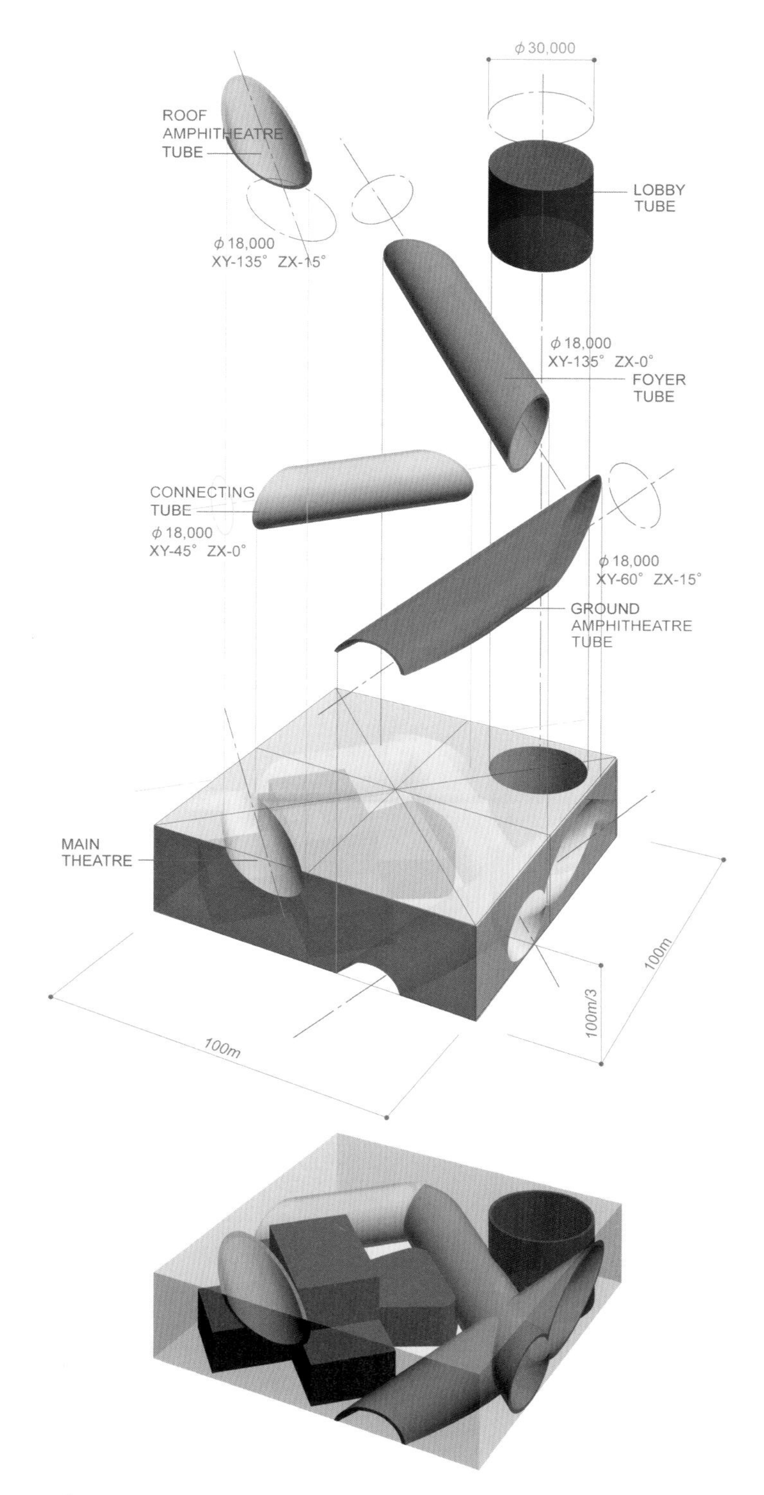

Diagram of the axis angles and diameters of the five cylinders inserted into cuboid
五个圆筒插入长方体的轴线角度和直径位置示意图

基地位于中国正蓬勃发展的上海市郊嘉定区。在建中的嘉定新城面积达到 10,000 公顷，该建筑作为新城的文化中心，是包括歌剧院在内的文化综合设施。

上海被称为“东方之珠”，自 19 世纪中叶港口开放以来，便以惊人的速度发展成为一座国际大都市，以及向全中国传播当代城市文明的文化基地。上海能有如此辉煌夺目的成就也许应该归功于“东方和西方”“传统和现代”这些固有的差异所碰擦出的火花。激情的都市也赋予这里的建筑极强的象征性和造型性。因此，我们渴望能在这里建造出可以通过建筑的内部空间（而非建筑的外形）来反映城市激情的建筑。

上海保利大剧院位于嘉定新城以东的公园中心，这里拥有宽广的人工湖和丰富的绿色植被。剧院正对人工湖，从外观上看，是一个底面为 100 米 x100 米，高 34 米的单纯长方体。在这一体量中，从上下左右各个角度插入了直径为 18 米的圆筒形空间，由此自然展开的内部三维空间成为了剧院的结构骨架，中央设置约 1,600 个座位的主会场。实与虚、长方体结构与筒状结构的相互交错，创造出各式各样具有特质的空间，在单纯的框架中，空间复杂而又千变万化。

在各个方向上延伸的圆筒，在建筑表皮与包裹长方体的幕墙相碰撞，由于碰撞角度的关系，从而形成了椭圆的开口部。这种大胆的曲线形式也暗示了在静谧的表面之下潜藏着的空间的剧烈性。

安藤忠雄建筑研究所 / 文

Opposite: Local residents participating in various activities at the semi-outdoor cylindrical space. This page: The discussion manuscript of the cylinder space. The entrance hall's diameter is finally determined to 30m in diameter and the rest is 18m in diameter. pp. 224–225: Light illuminates the entrance hall's horizontally-aligned wood grille, whose contrast with the fair-faced concrete and stairs creates layers of meaning in the space.

对页：市民在剧院半室外圆筒空间内活动的场景。本页：圆筒空间的研讨手稿，最终确定入口大厅为直径 30 米，其余直径 18 米。第 224-225 页：大厅天井的采光照亮了整个横向木格栅饰面墙体，搭配上清水混凝土墙面以及阶梯，创造出丰富的层次感。

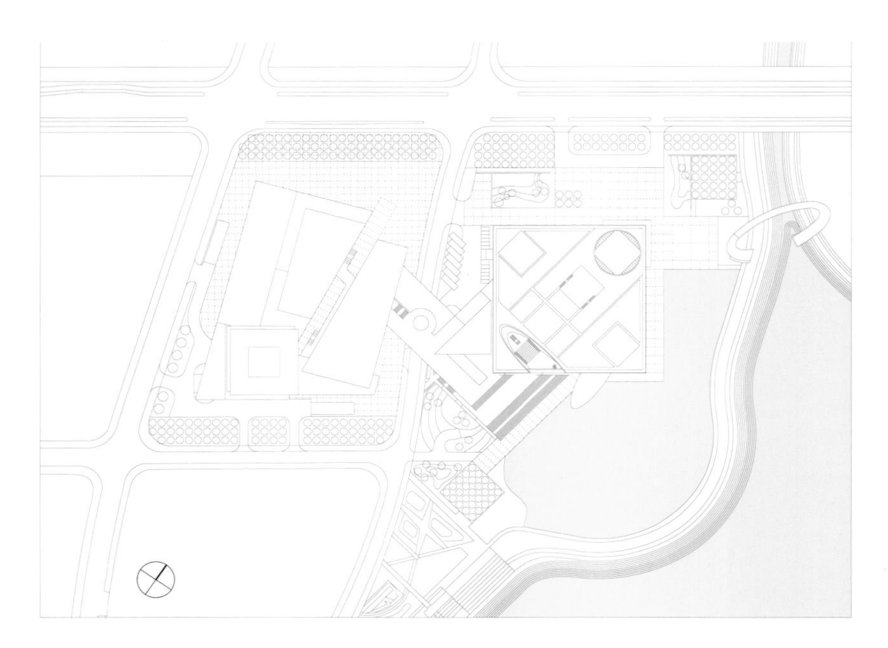

Site plan (scale: 1/5,000)／总平面图（1/5,000）

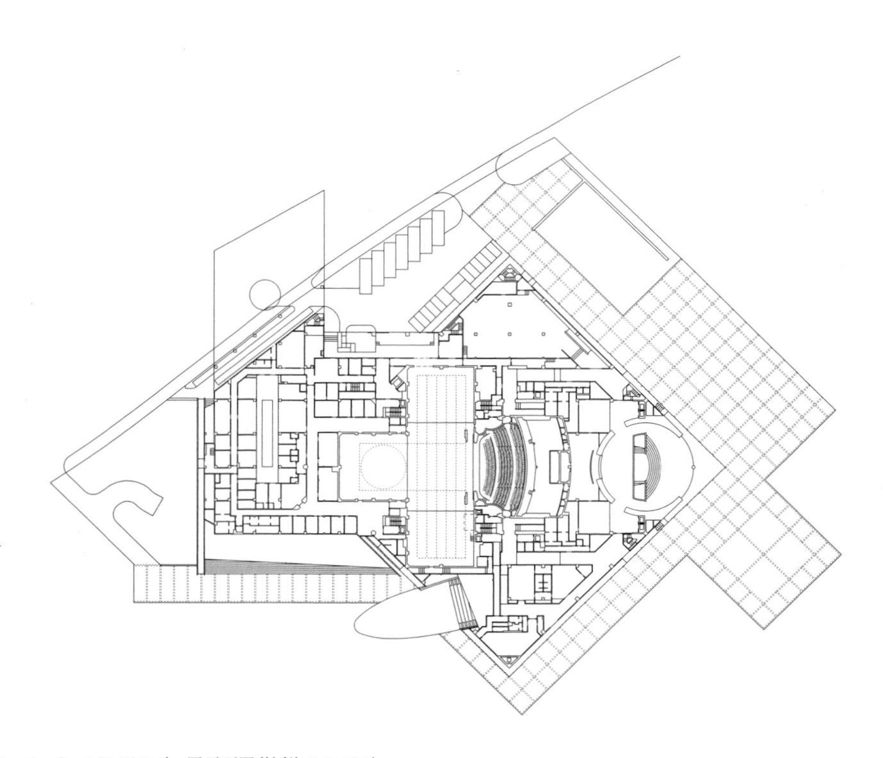

First floor plan (scale: 1/2,500)／一层平面图（比例：1/2,500）

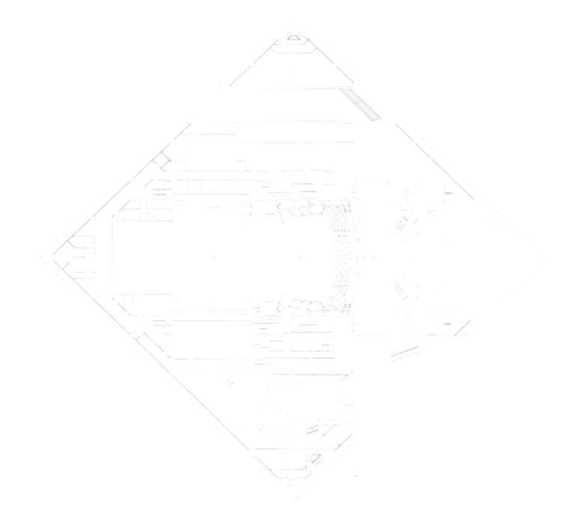

Fourth floor plan ／四层平面图

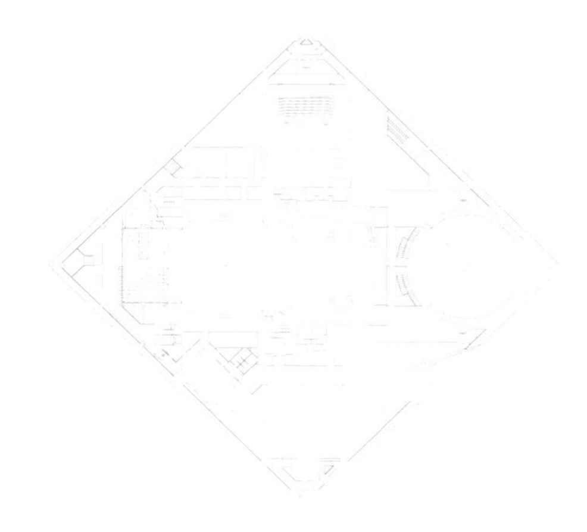

Fifth floor plan ／五层平面图

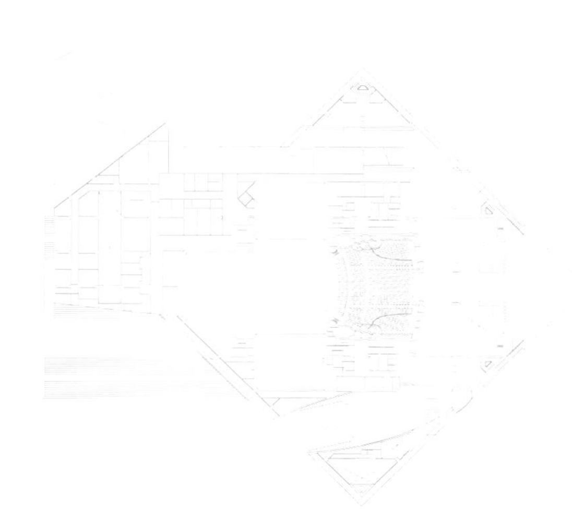

Second floor plan ／二层平面图

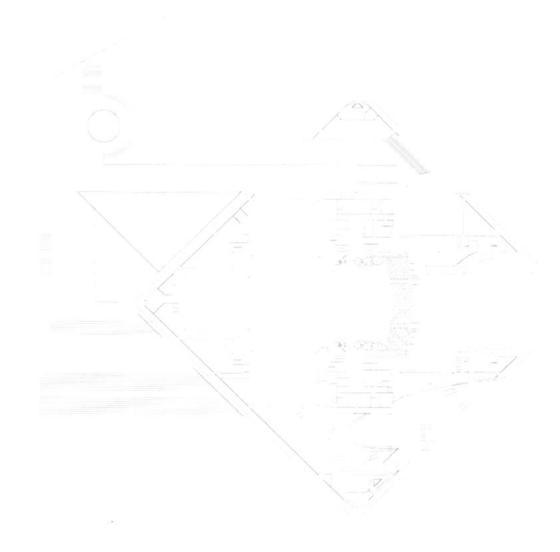

Third floor plan／三层平面图

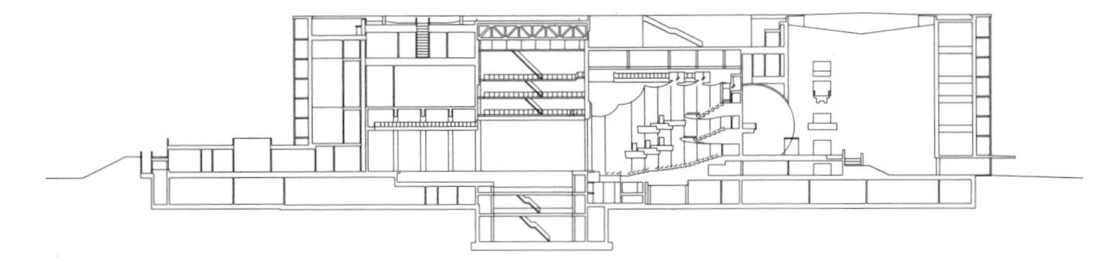

Section A-A／剖面图A-A

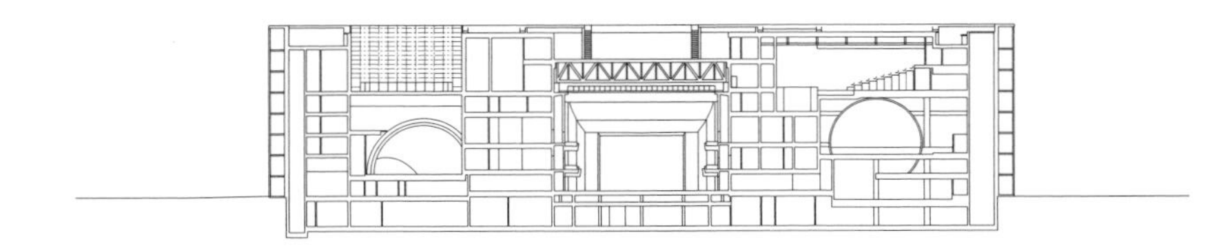

Section B-B／剖面图B-B

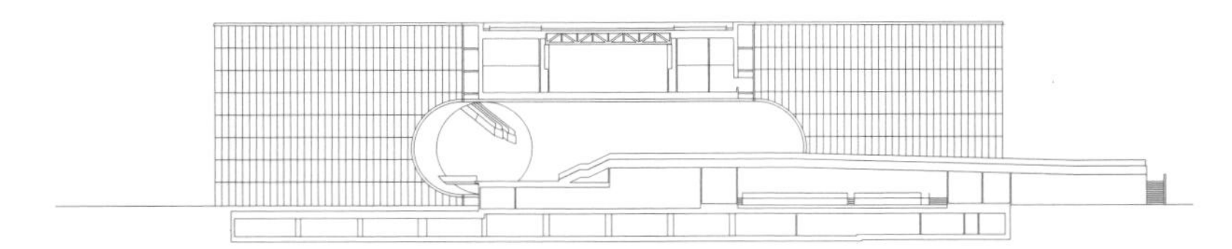

Section C-C／剖面图C-C

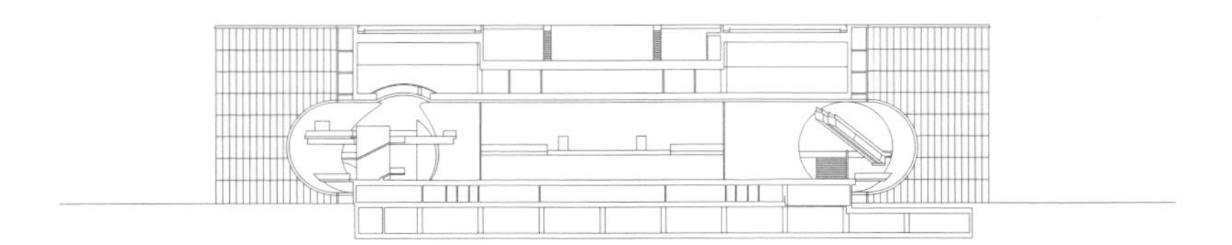

Section D-D／剖面图D-D

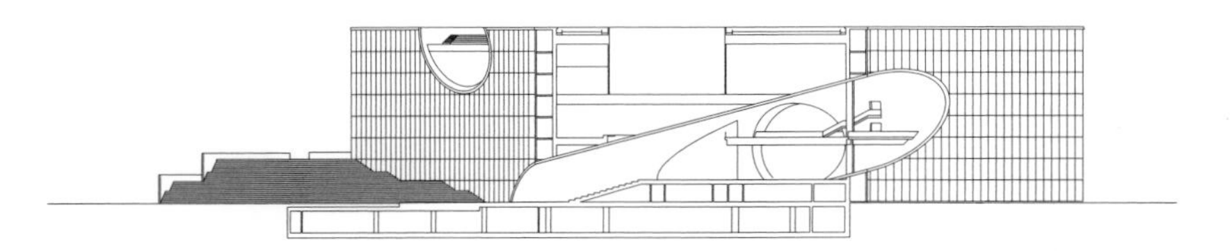

Section E-E／剖面图E-E

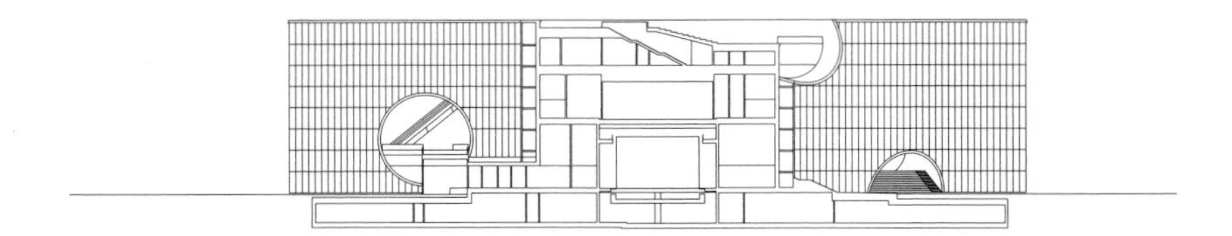

Section F-F (scale: 1/2,000)／剖面图F-F（比例：1/2,000）

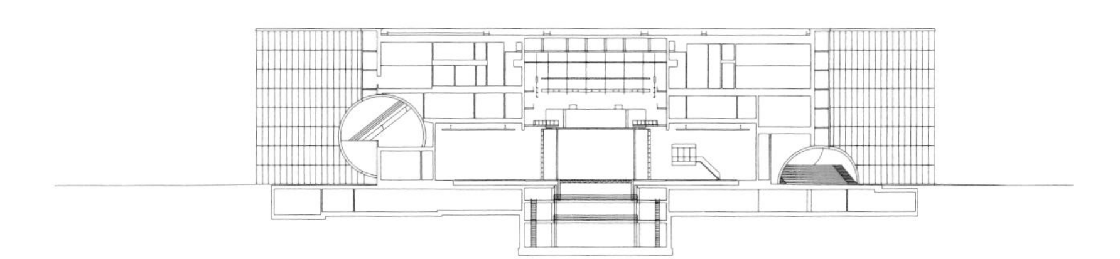

Section G-G／剖面图G-G

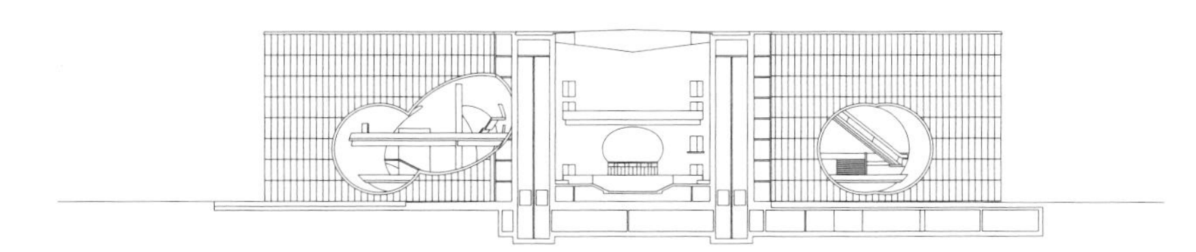

Section H-H／剖面图H-H

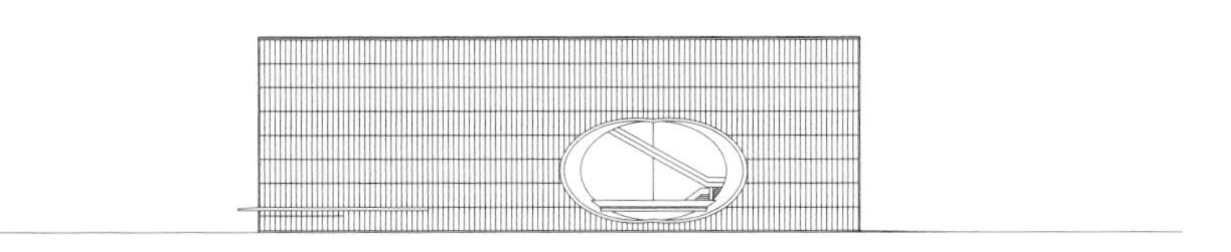

North elevation／北立面图

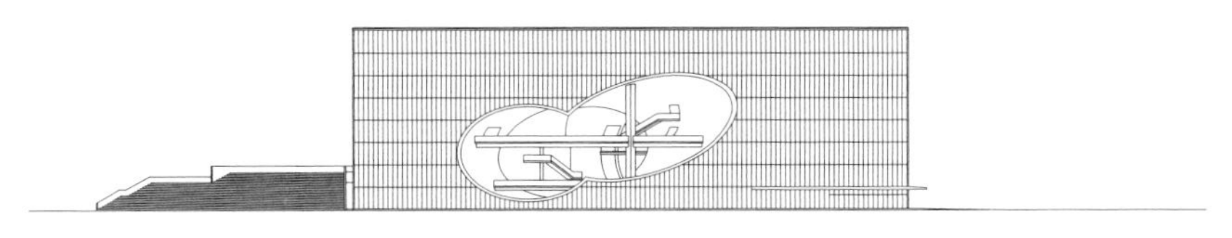

East elevation／东立面图

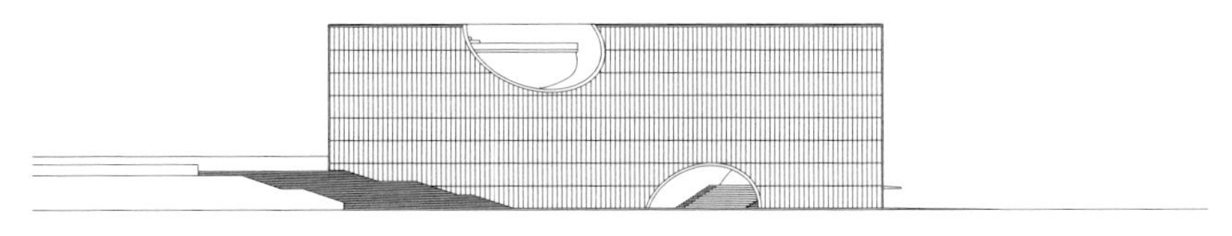

South elevation／南立面图

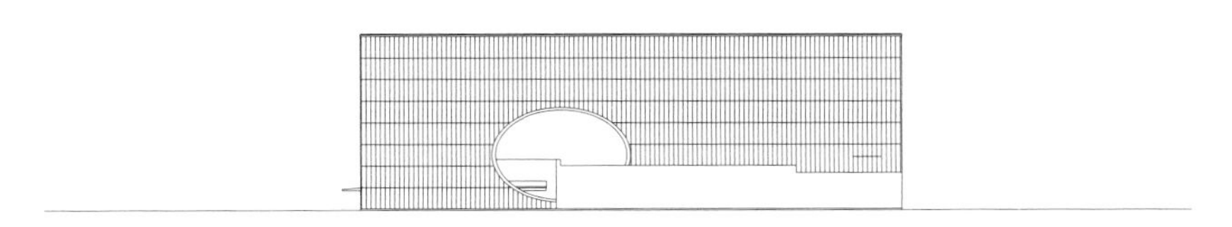

West elevation (scale: 1/2,000)／西立面图(比例:1/2,000)

B 入口

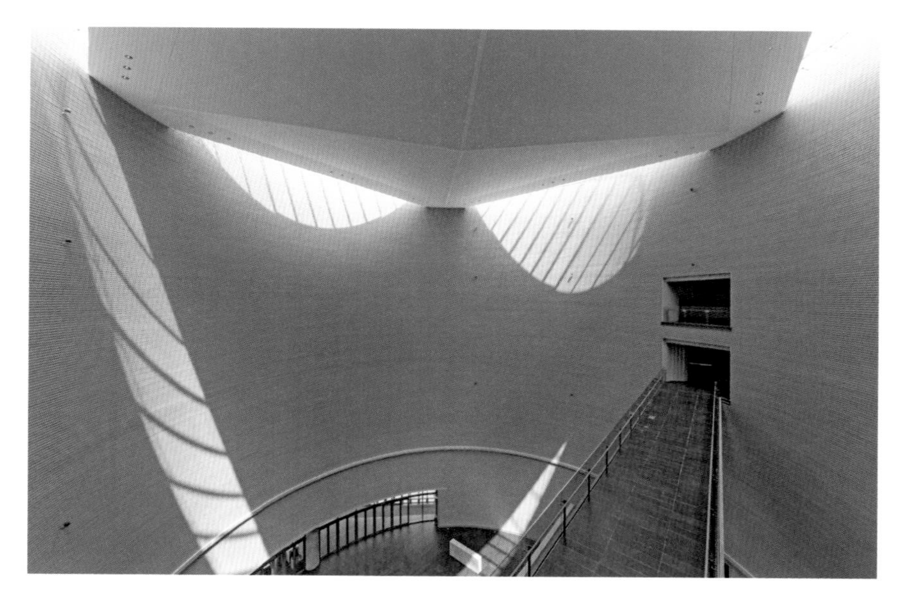

Opposite, above: The 18-meter-wide, 90-meter-long tubular foyer intersects the central axis that extends from the entrance to the theater. It holds dynamic views that pierce throuth the east facade on the lake side and the front facade on the Baiyin Road side. Opposite, below: Interior of the theater. This page, above: Light from overhead adds expression to the dynamic cylindrical space that extends cross five stories. This page, below:Visitors are greeted by a cylindrical entrance lobby measuring 30 meters in height and width. All images on pp. 216–229 by Shinkenchiku-sha except the specified.

对页，上：与建筑中轴正交的从入口向剧院、直径 18 米、全长 90 米的圆筒状休息大厅。贯穿远香湖一侧的东立面到白银路一侧的正立面的生动视角；对页，下：剧院内景。本页，上：五层挑空的圆筒空间内，从上洒下的光富裕了内部丰富的空间表情；本页，下：迎接来访者的直径 30 米、高 30 米的圆筒状入口大厅。